BASIC SKILLS WITH MATH

BASIC SKILLS WITH MATH

Geometry

CONNIE EICHHORN

CAMBRIDGE ADULT EDUCATION
Upper Saddle River, New Jersey
www.globefearon.com/welcome/cambridge.html

EDITORIAL DEVELOPER: Cathy Fillmore Hoyt

EDITORS: Stephanie Cahill, Doug Falk, Dena Pollak, Phyllis Dunsay

PRODUCTION EDITOR: Alan Dalgleish, Suzanne Keezer

BOOK DESIGN: Parallelogram, New York

ELECTRONIC PAGE PRODUCTION: Burmar Technical Corporation, Albertson, New York

COVER ART: Salem Krieger

COVER DESIGN: Patricia Battipede

Printed in the United States of America

2 3 4 5 6 7 8 9 10 02 01 00

ISBN 0-835-95729-2

CAMBRIDGE ADULT EDUCATION
Upper Saddle River, New Jersey
www.globefearon.com/welcome/cambridge.html

Contents

Unit 5. Solids

Unit 6. The Coordinate Plane

GEOMETRY PREVIEW

The problems on pages 1 through 5 will help you find out which parts of this book you need to work on. Do all the problems you can. At the end of the problems, look at the chart to see which page you should go to next.

Name the kind of angle.

1. 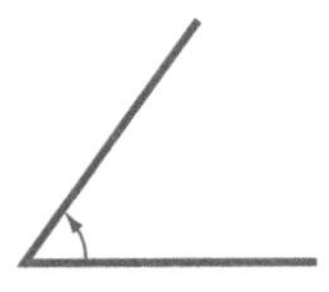

____

2. 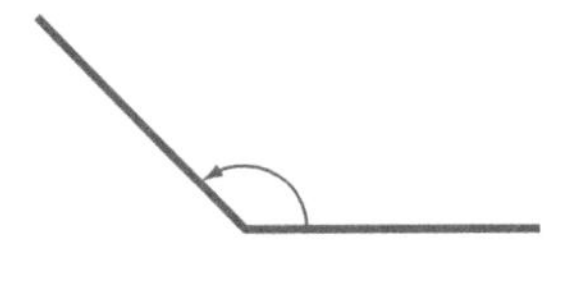

____

3. 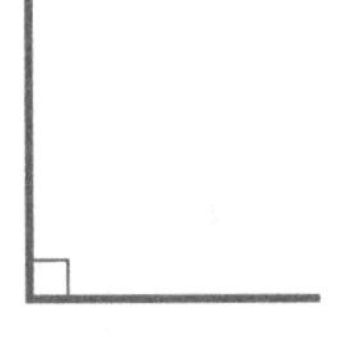

____

4. 

____

5. 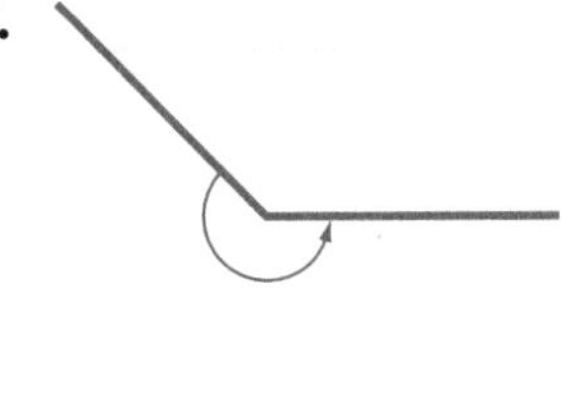

____

6. What is the complement of a 37° angle? ____________

7. What is the supplement of a 48° angle? ____________

Use the figure to answer items 8–13.

N
1 2
L
3 4
5 6
M
7 8

8. What is the corresponding angle for $\angle 1$? ____________

9. What type of lines are L and M? ____________

10. What type of line is N? ____________

11. Which angle is vertical to $\angle 8$? ____________

12. If $\angle 7$ is 65°, what is the measure of $\angle 5$? ____________

13. If $\angle 7$ is 65°, what is the measure of $\angle 8$? ____________

Name the type of triangle.

14.

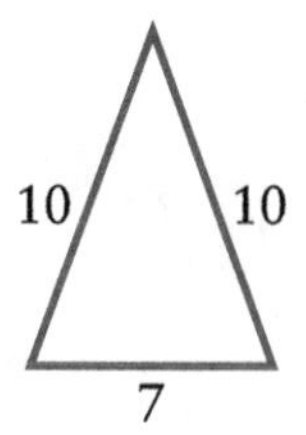

____

15.

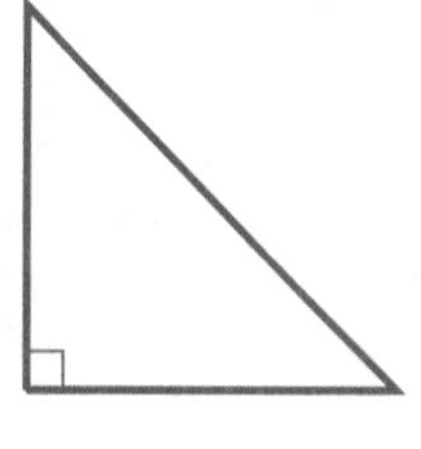

____

16.

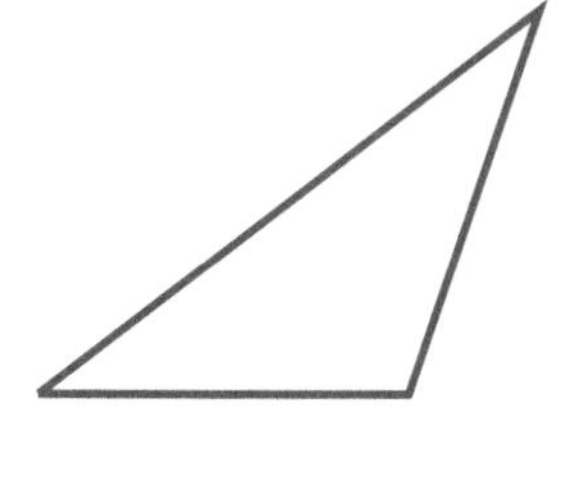

____

17.

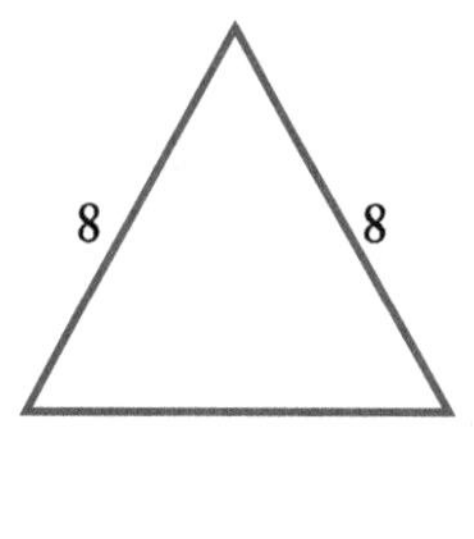

____

Find the perimeter and area of the triangle below.

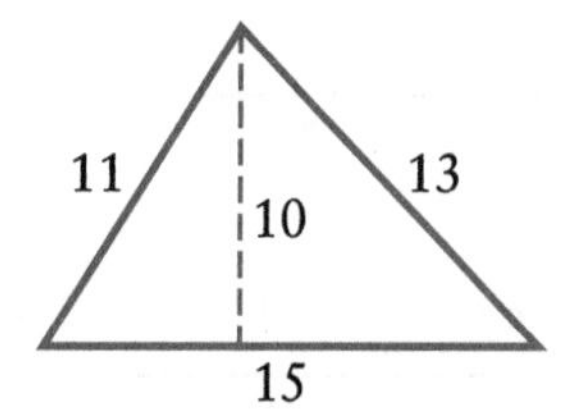

18. $P =$ ____________ **19.** $A =$ ____________

Find the length of the unknown side.

20.

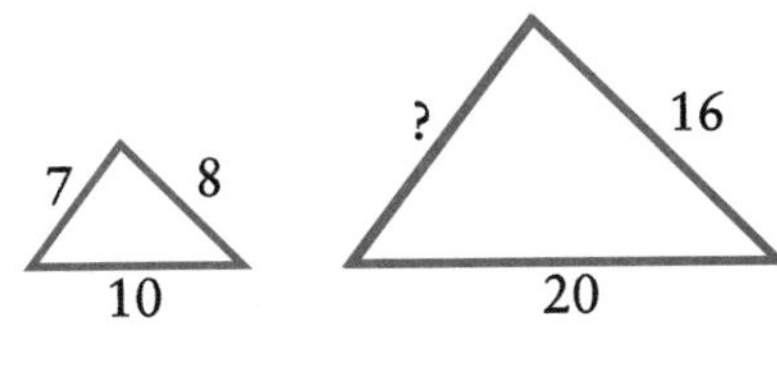

similar triangles

____

21.

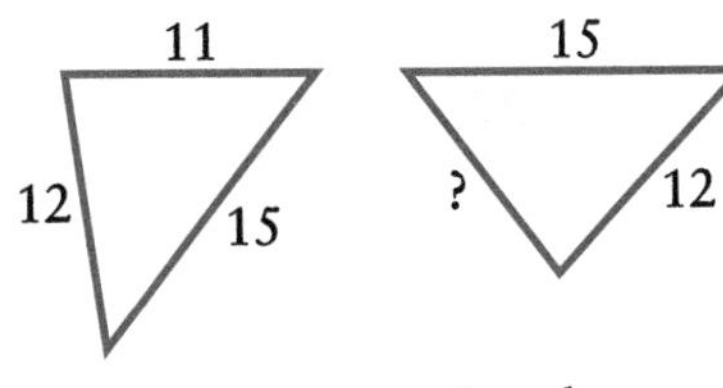

congruent triangles

____

Name the kind of quadrilateral.

22.

____

23.

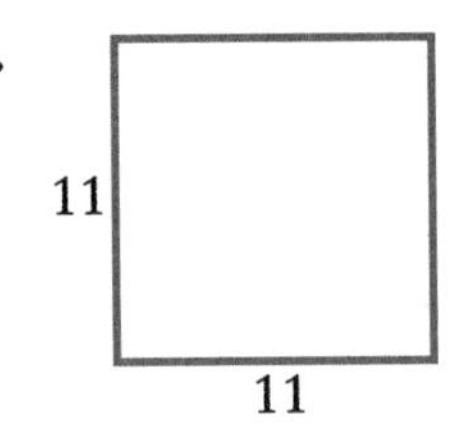

____

24.

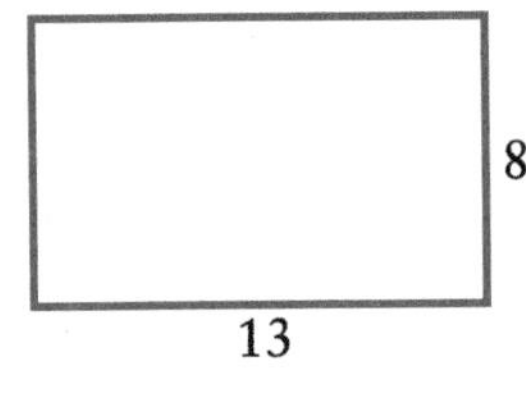

____

25.

____

Find the area and perimeter (or circumference) of each. Use 3.14 for π.

26.

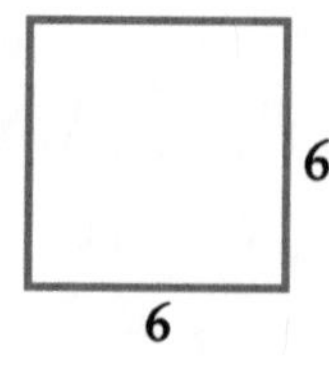

____

27.

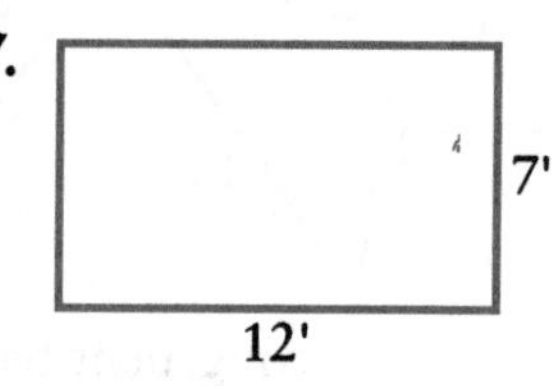

____

28.

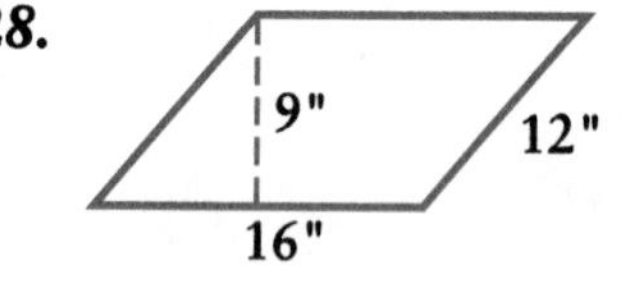

____

29.

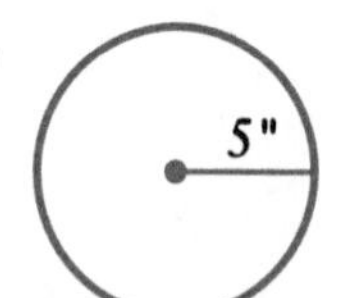

____

30.

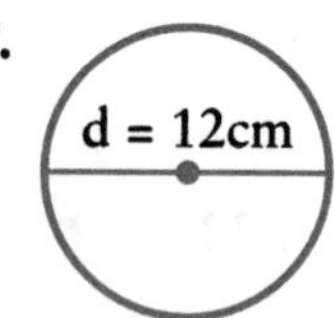

____

Name the type of solid.

31.

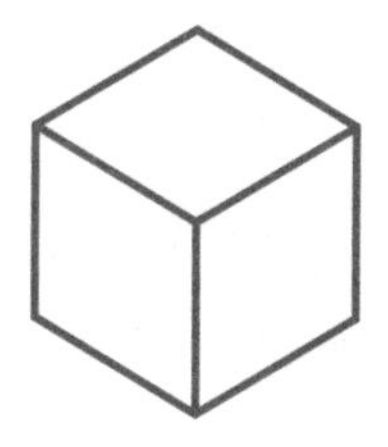

____

32.

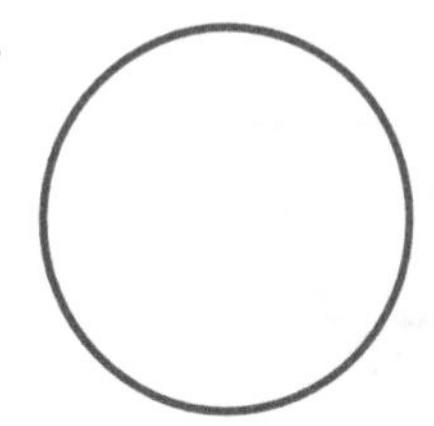

____

33.

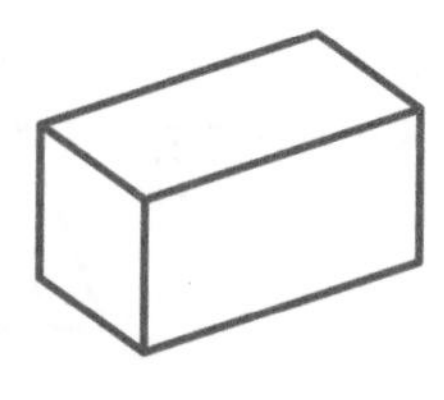

____

34.

____

35.

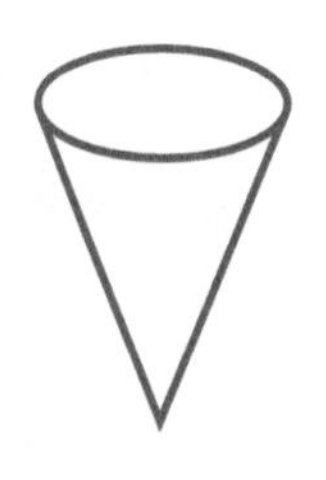

____

Find the volume of the following solids.

36.

$r = 5"$
$h = 4"$

V = ________

37.

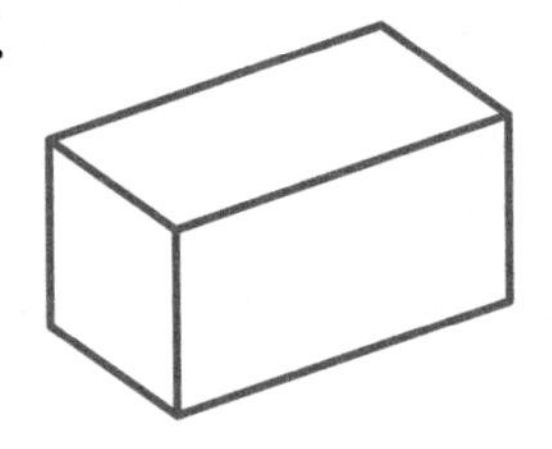

$l = 12"$
$w = 6"$
$h = 10"$

V = ________

38.

$s = 8cm$

V = ________

Write the coordinates for the points on the graph.

39. *A*

40. *B*

41. *C*

42. *D*

43. *E*

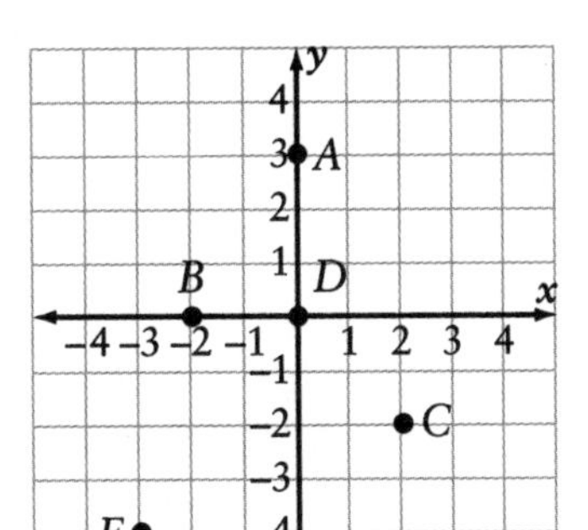

PROGRESS CHECK

Check your answers on page 111. Then complete the chart below.

Problem numbers	*Number of problems in this section*	*Number of problems you got right in this section*	
1 to 13	13	______________	If you had fewer than 10 problems right, go to page 7.
14 to 21	8	______________	If you had fewer than 6 problems right, go to page 29.
22 to 30	9	______________	If you had fewer than 7 problems right, go to page 54.
31 to 38	8	______________	If you had fewer than 6 problems right, go to page 79.
39 to 43	5	______________	If you had fewer than 4 problems right, go to page 96.

UNIT

I

Angles and Lines

Defining an Angle

You see and work with angles every day. The corner of a building, the slant of a roof, and the opening of scissors all form angles. When two straight lines meet at a point, they form an **angle.** Each of the straight lines is called a **side** of the angle.

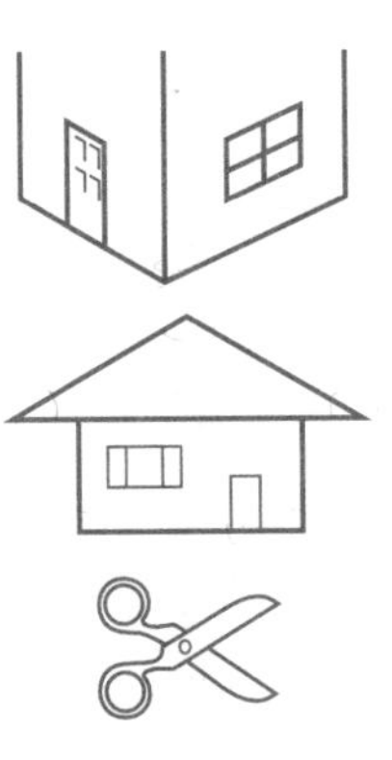

Look at the angles below. In the angle at the left, the end of each side and the vertex have been given a letter, A, B, or C. The point where the lines meet is the **vertex** and is given the middle letter, in this case, B. The curved line with the arrow shows the angle, called $\angle ABC$ or $\angle CBA$.

You can also name an angle by labeling only the vertex. To refer to the middle angle, you would say $\angle B$.

A third way to name an angle is to use the number inside the angle. This angle on the right is $\angle 2$.

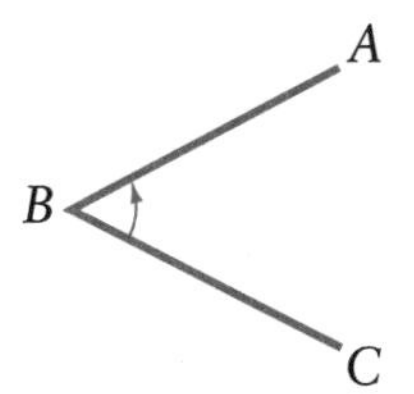

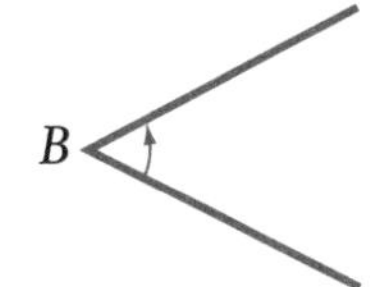

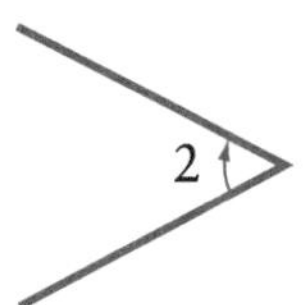

Example: Label each of the angles in a square.

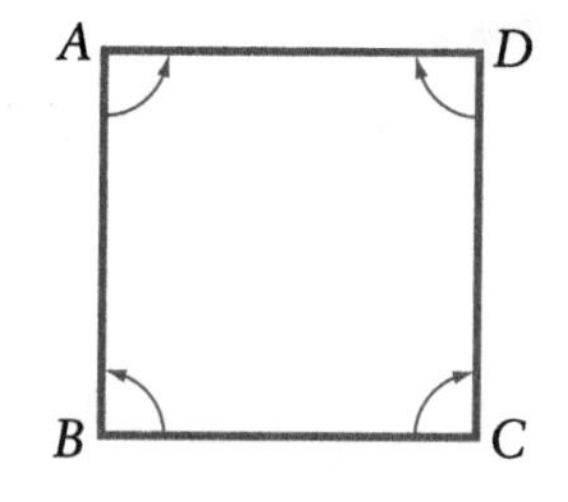

Step 1. Put a letter at the end of each straight line.

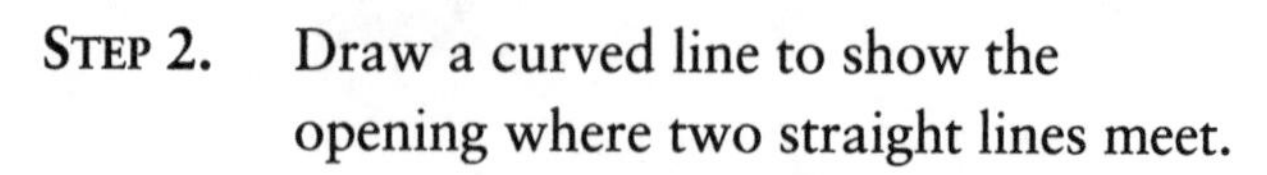

Step 2. Draw a curved line to show the opening where two straight lines meet.

Step 3. Label the angles with the letter of the lines that form the angle. Use the middle letter as the vertex.

➠ There are four angles in the square: $\angle ABC$, $\angle BCD$, $\angle CDA$, and $\angle DAB$ or $\angle B$, $\angle C$, $\angle D$, and $\angle A$.

PRACTICE 1

Write all the possible labels for each angle.

1.

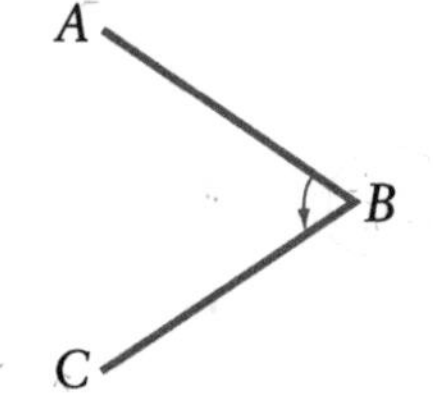

2.

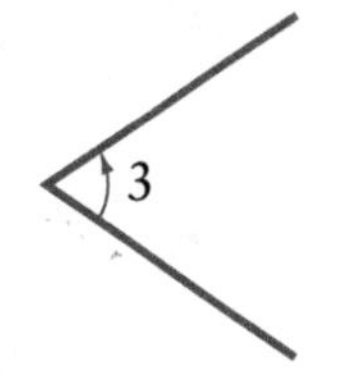

3.

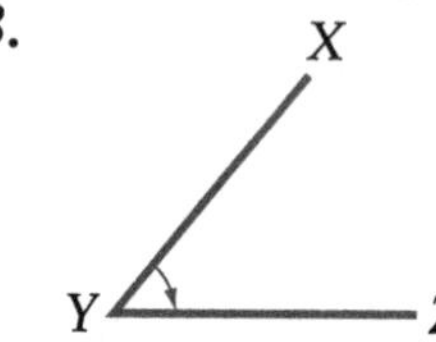

4.

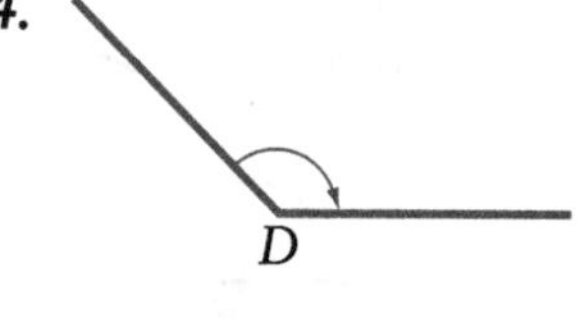

5.

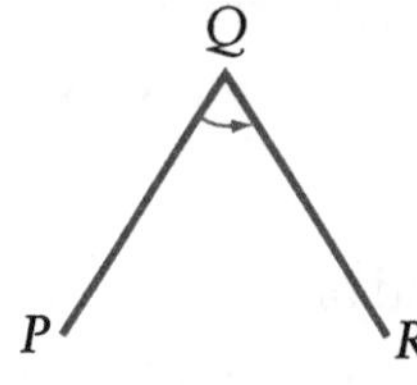

6.

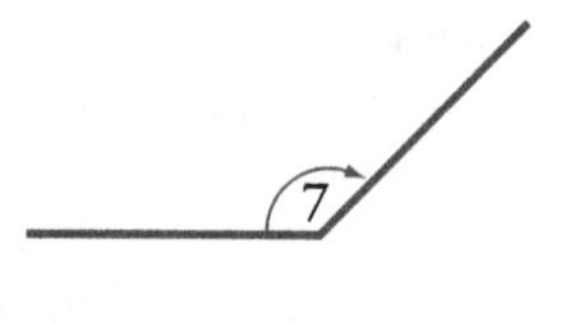

Draw and label as directed.

7. Draw an angle. Label it $\angle 3$.

8. Draw an angle. Label it $\angle DEF$.

9. Draw an angle. Label it $\angle C$.

Match the angle to its name.

10.

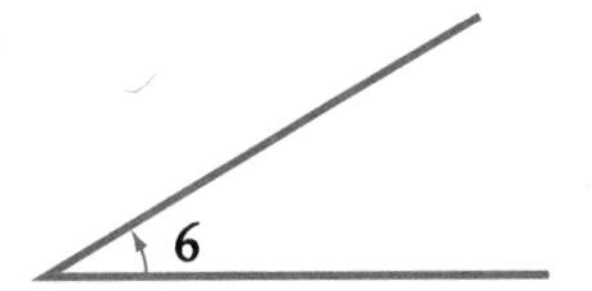

a. $\angle PQR$

b. $\angle TSR$

c. $\angle M$

d. $\angle 6$

e. $\angle A$

11.

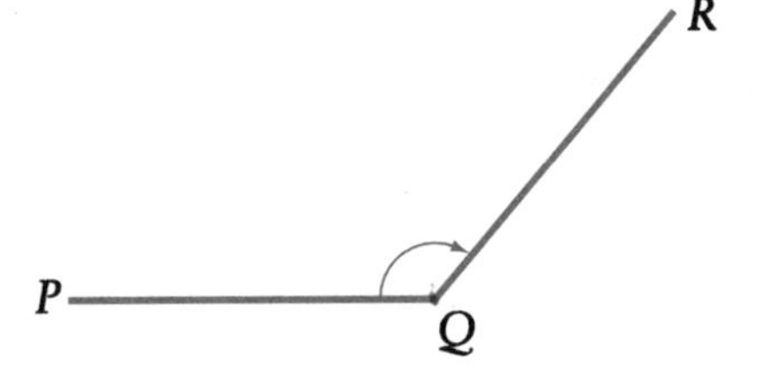

12.

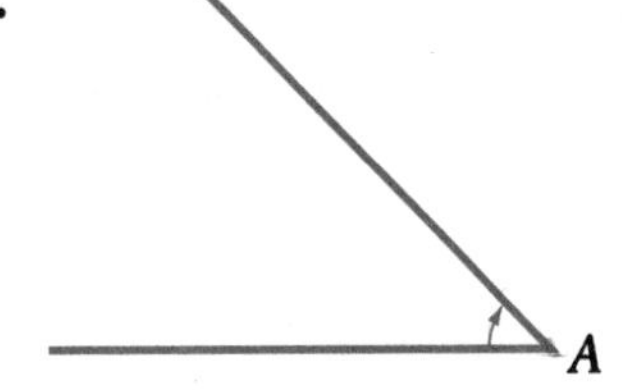

13.

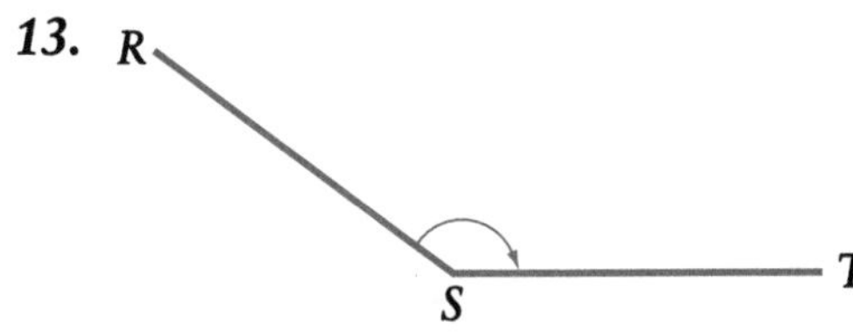

14.

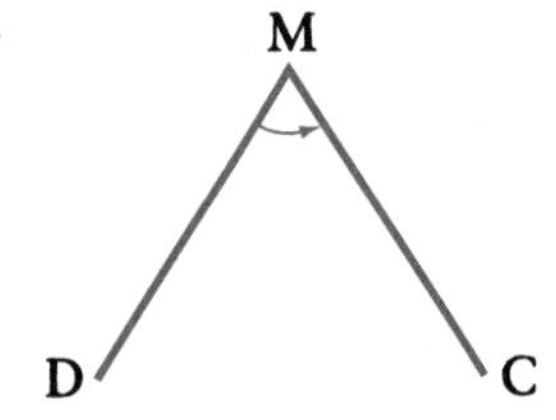

Define the terms in items 15 and 16.

15. vertex ______________________________

16. angle ______________________________

17. Name three places where you see angles every day in your home.

To check your answers, turn to page 111.

Using a Protractor

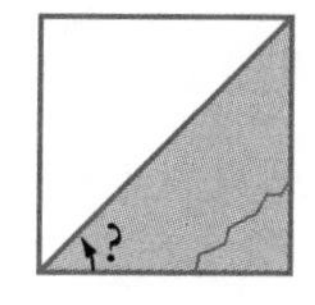

Carlos is fixing the bathroom floor. One of the tiles is broken and must be replaced. To cut the new tile to fit, Carlos must measure the angle of the broken piece. How many degrees is the angle?

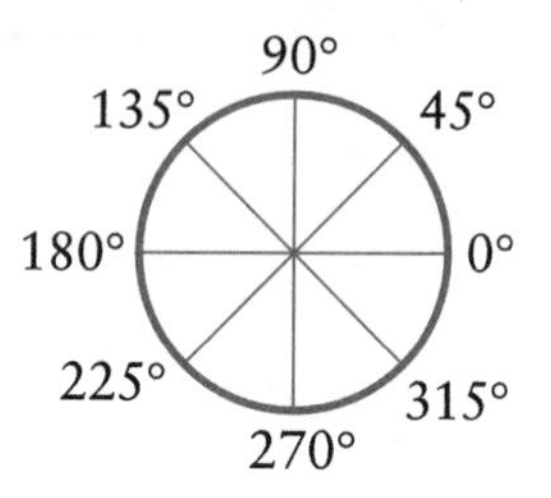

Angles are measured in units called **degrees.** ° is the symbol for degree. A degree is part of a circle. The whole circle is 360°. The figure at the right shows how a circle divides up into degrees.

Angles can be measured using a **protractor.** The protractor is the shape of a half circle. The bottom is a straight line with a hole or mark at the exact center, just above the bottom edge. The curved top of the protractor is marked off in degrees—from 0° to 180° on the outer edge, and from 180° to 0° on the inner edge.

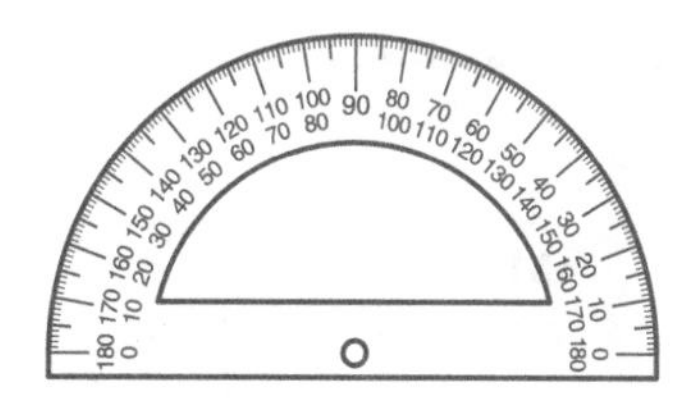

Example: Measure the degrees of the angle of the broken tile.

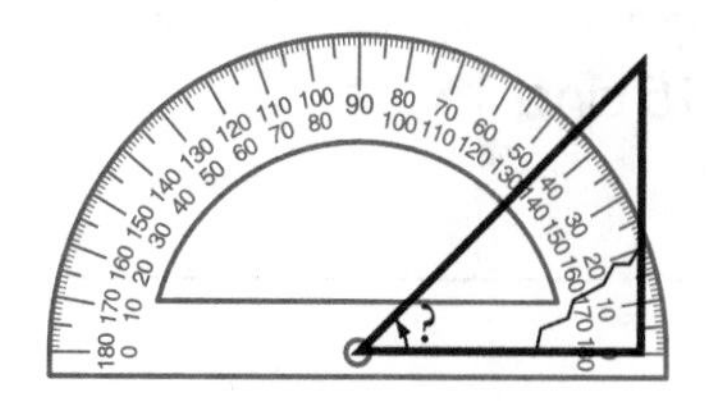

Step 1. Place the center hole or mark of the protractor over the vertex.

Step 2. Line up the straight edge of the protractor with the lower side of the angle.

Step 3. Read the number of degrees where the other side of the angle crosses the circular edge of the protractor.

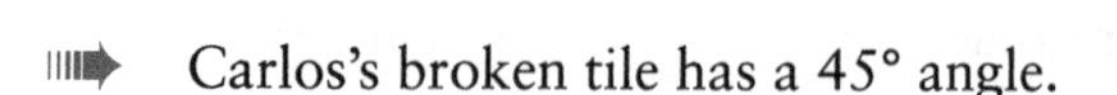

Carlos's broken tile has a 45° angle.

PRACTICE 2

Write the measure of each angle.

1.

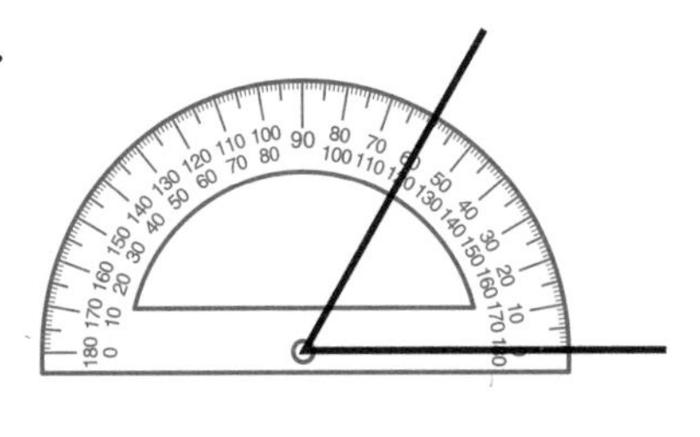

____

2.

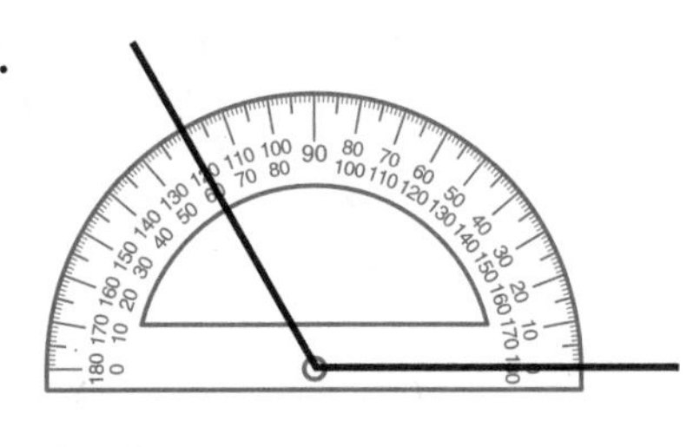

____

3.

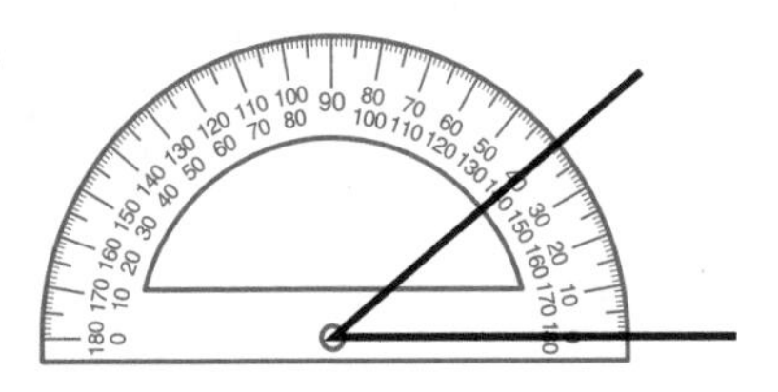

____

4.

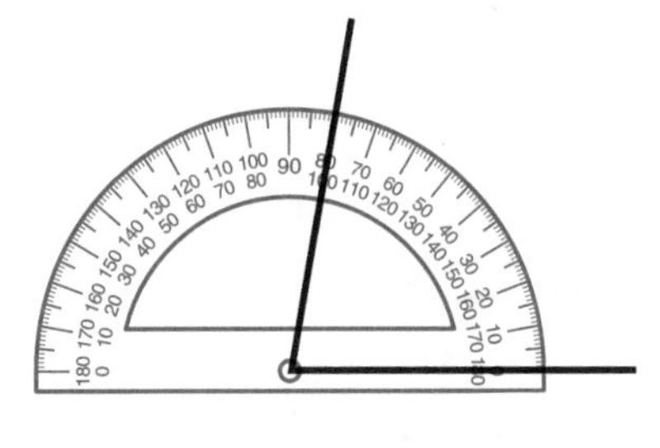

____

5.

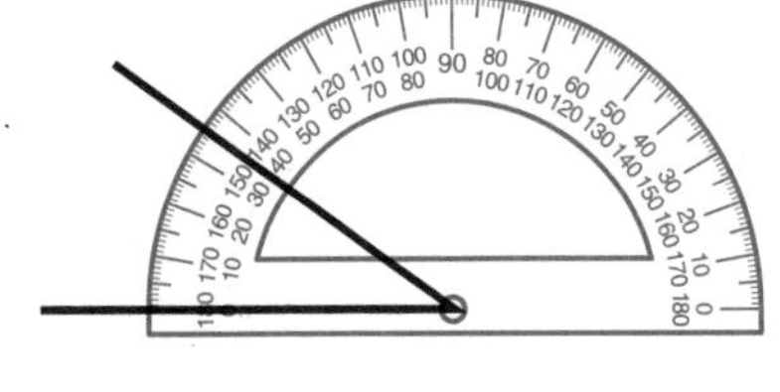

____

6.

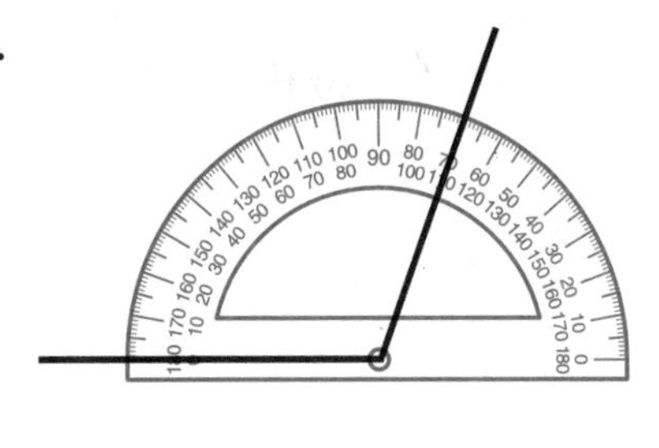

____

7.

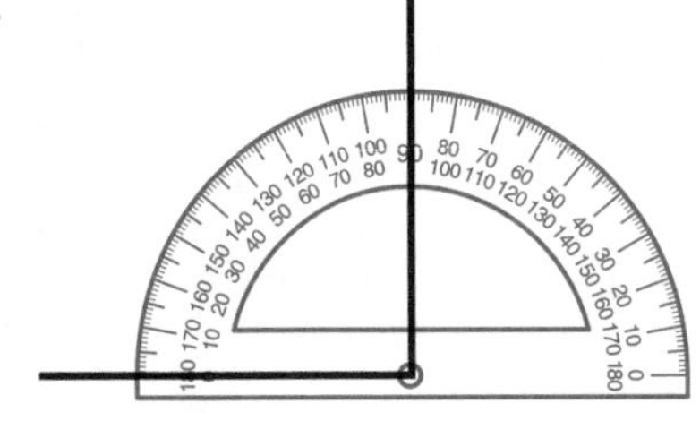

____

8.

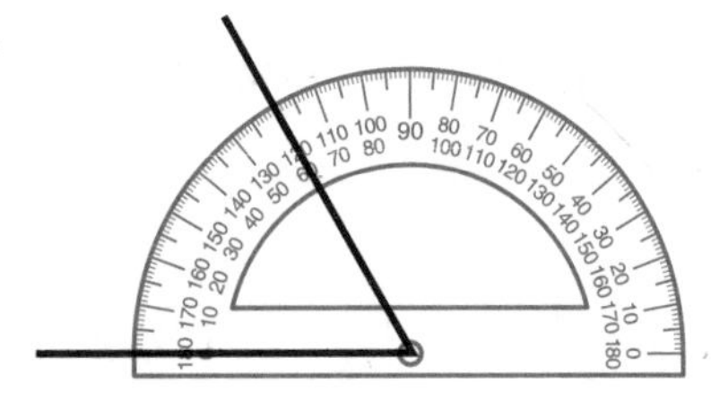

____

Use a protractor to find the measure of each angle.

9.

10.

11.

12.

13. Freda is making a mosaic out of small tile pieces. Here is a diagram of an angle she must cut. What is the measure of the angle?

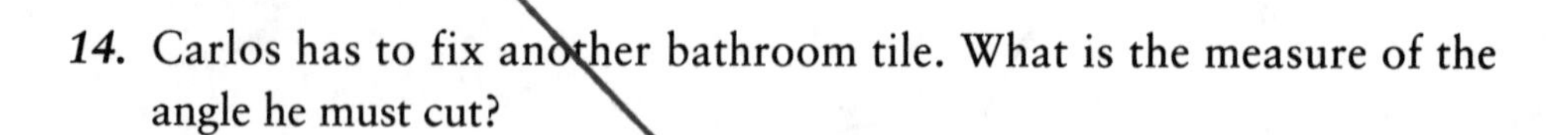

14. Carlos has to fix another bathroom tile. What is the measure of the angle he must cut?

To check your answers, turn to page 112.

Kinds of Angles

The size of an angle determines the kind of angle it is. Here are five different kinds of angles. An **acute angle** is more than 0° but less than 90°. A **right angle** is exactly 90°.

An angle more than 90° but less than 180° is called an **obtuse angle. A straight angle** is exactly 180°. If the angle is more than 180° but less than 360°, it is a **reflex angle.**

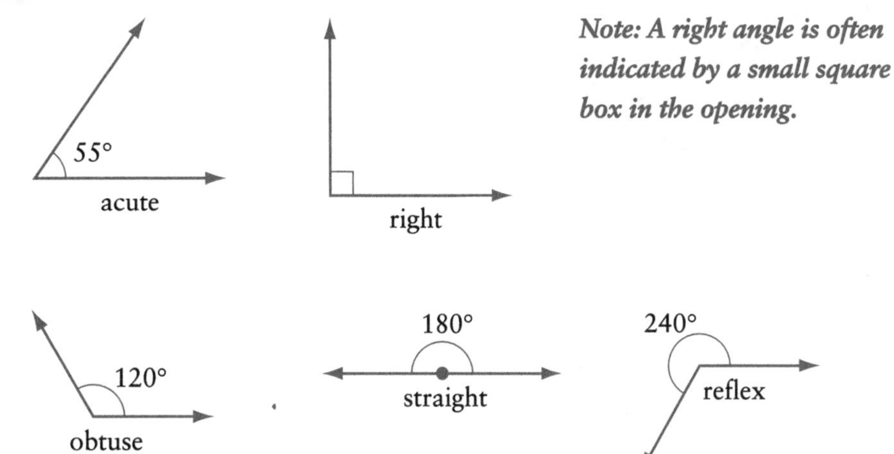

Note: A right angle is often indicated by a small square box in the opening.

Example: Find what kind of angle is formed in the illustration.

STEP 1. With a protractor, find the number of degrees in the angle.

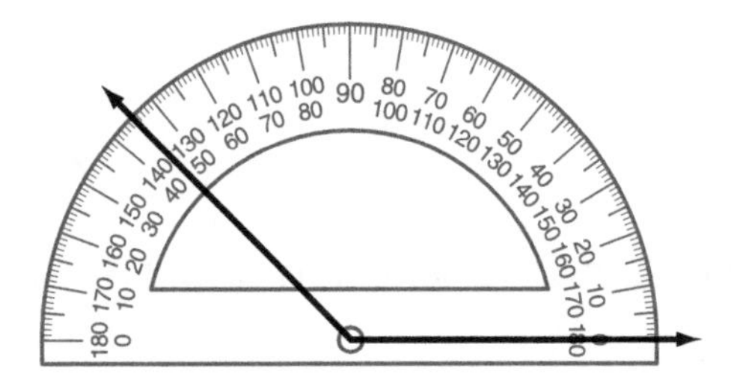

STEP 2. Find the kind of angle, based on the table below:

Number of degrees	*Kind of Angle*
more than 0°, less than 90°	Acute
exactly 90°	Right
more than 90°, less than 180°	Obtuse
exactly 180°	Straight
more than 180°, less than 360°	Reflex

➡ The angle formed is 135°. It is an obtuse angle.

PRACTICE 3

What kind of angle is each of the following?

1\. 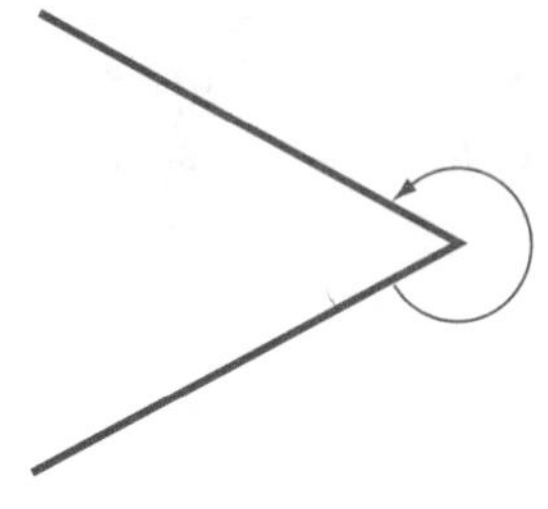

____

2\. 

____

3\. 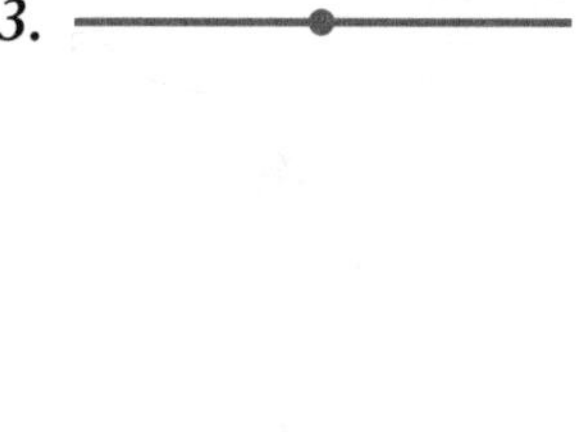

____

4\. 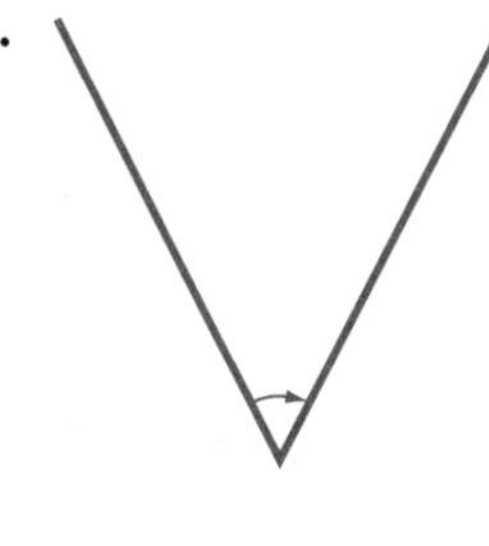

____

5\. 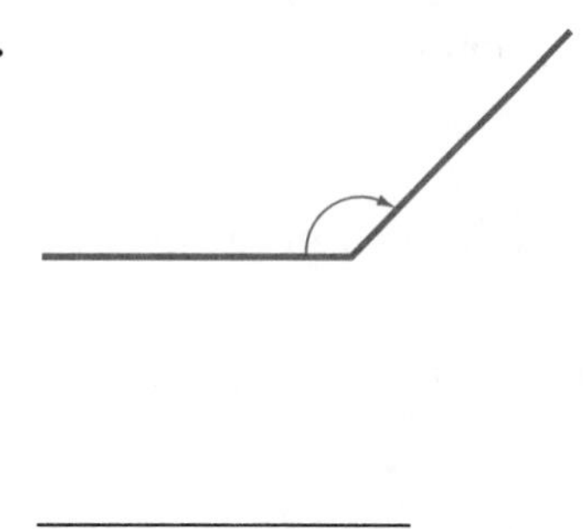

____

Draw two examples of each kind of angle. Then label each angle using letters or numbers.

6\. acute

7\. right

8\. obtuse

9\. straight

10\. reflex

Match.

_______ *11.* acute *a.* 90°

_______ *12.* straight *b.* less than 90°

_______ *13.* obtuse *c.* 180°

_______ *14.* right *d.* greater than 90°

_______ *15.* reflex *e.* greater than 180°

16. On a road map, Kina saw two streets that intersected as shown in the diagram. What type of angle do the streets form?

To check your answers, turn to page 112.

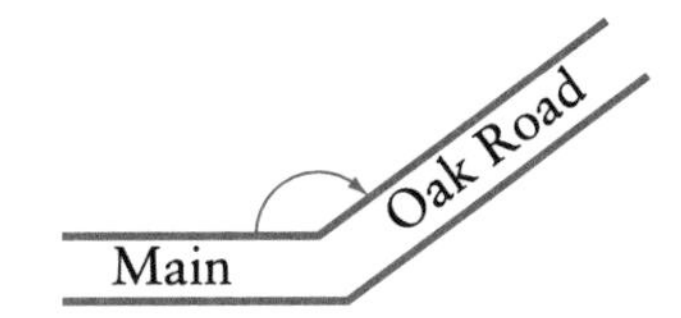

Complementary Angles

Jill is building a picture frame. Each corner of the frame needs to be a right angle. One piece of wood has been cut at 45°. At what angle should she cut the second piece?

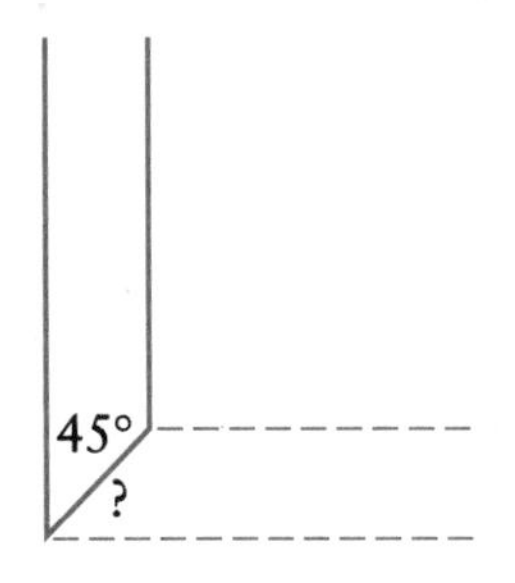

Angles can be added together to make a single, larger angle. When two angles total 90°, or a right angle, they are called **complementary angles.** One angle is the complement of the other.

Example: Find the complement of the 45° piece of wood Jill needs to cut.

Step 1. Find the number of degrees in the first angle. 45°

Step 2. Subtract the number of degrees from 90°. 90° − 45° = 45°

➡ Jill needs to cut the second piece at a 45° angle to make the two pieces complementary, to form a right angle.

PRACTICE 4

Find the complement of each angle.

1. 60° ________ *2.* 3° ________ *3.* 80° ________

4. 70° ________ *5.* 26° ________ *6.* 89° ________

Measure each angle with a protractor. Then find its complement.

7.

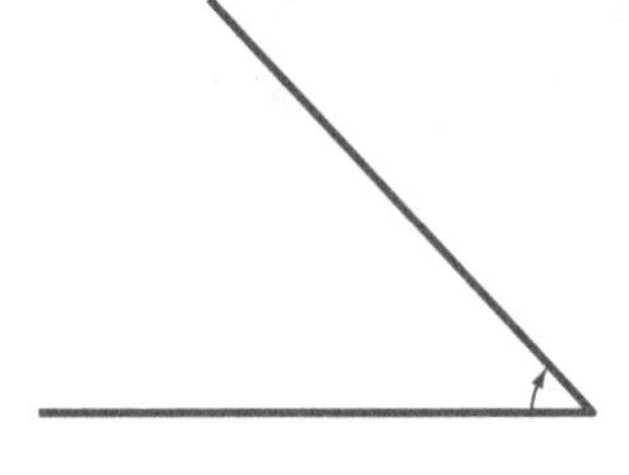

Angle ________

Complement ________

8.

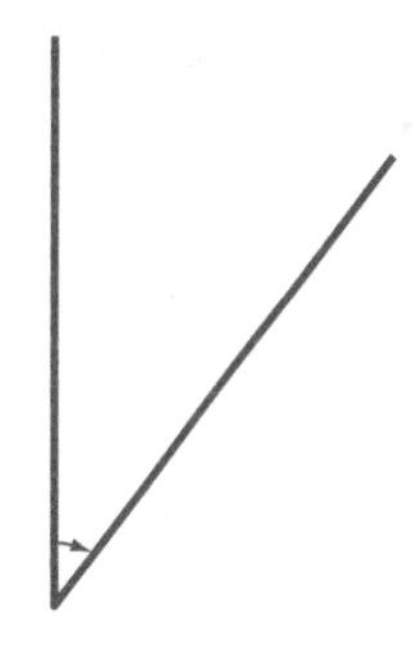

Angle ________

Complement ________

9.

Angle ________

Complement ________

10.

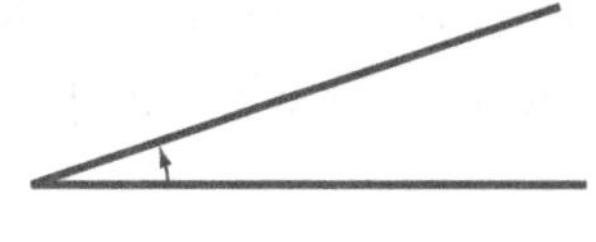

Angle ________

Complement ________

11. Grace is making a pillow cover from small pieces of fabric. Each corner should make a right angle. If one piece of fabric has a 65° angle, at what angle should she cut the second piece to make a right angle?

To check your answers, turn to page 112.

Supplementary Angles

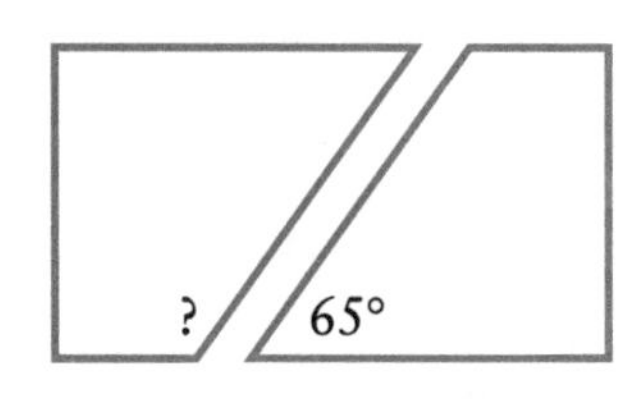

Maria is cutting out patterns for a quilt. The first piece she cuts has an angle of 65°. She wants to cut another piece so that the sides will make a straight line. At what angle must she cut the second piece?

When two angles added together total 180°, they are called **supplementary angles.** When the sides of supplementary angles are placed together, they form a straight line. One angle is the supplement of the other.

Example: If the first piece has a 65° angle, determine at what angle Maria should cut the second piece so that the bottom sides of the pieces form a straight line.

STEP 1. Find the number of degrees in the first angle. 65°

STEP 2. Subtract the number of degrees from 180°. $180° - 65° = 115°$

➡ Maria must cut the second piece at a 115° angle to make the second angle supplementary to the first.

PRACTICE 5

Find the supplement of each angle.

1. 40° ________ **2.** 72° ________ **3.** 135° ________

4. 120° ________ **5.** 97° ________ **6.** 5° ________

Measure each angle. Find the supplement of each.

7.

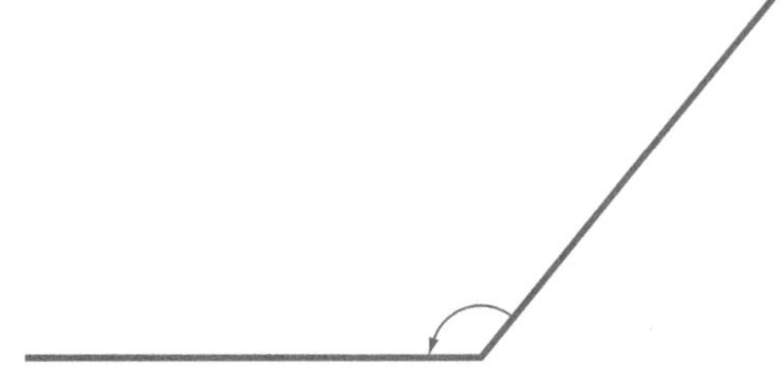

Angle ________

Supplement ________

8.

Angle ________

Supplement ________

9.

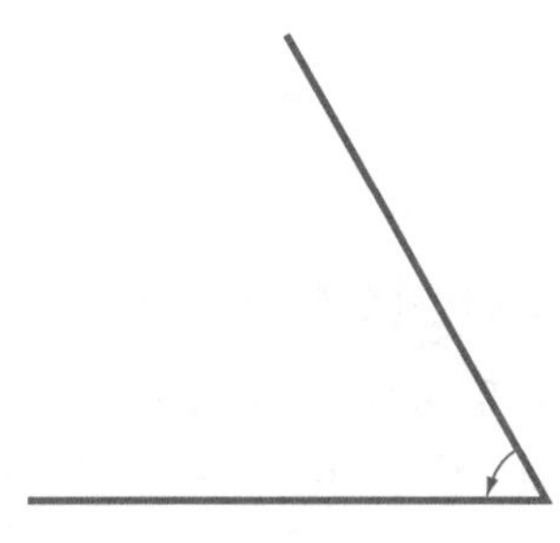

Angle ________

Supplement ________

Choose the supplement of each angle.

10. 46° *a.* 54° *b.* 46° *c.* 134°

11. 90° *a.* 90° *b.* 0° *c.* 45°

12. 130° *a.* 130° *b.* 50° *c.* 70°

13. 21° *a.* 21° *b.* 159° *c.* 69°

14. Angela was making a wooden puzzle out of geometric shapes. The shapes were to make a straight line when they fit together. If the angle is 100° on one piece, what angle must the fitting piece be?

15. Here is another puzzle piece. What is the measure of the piece that will be supplementary to it?

To check your answers, turn to page 113.

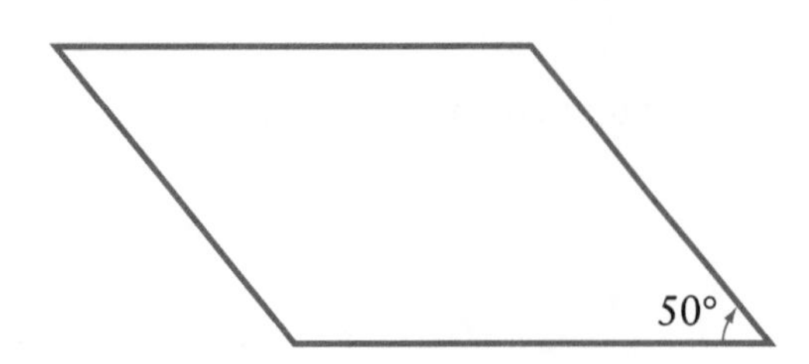

Parallel Lines

Bruce wants to put two 20-ft. boards a few inches apart. Rather than draw 20-ft. lines on the floor, he draws 1-ft. lines a few inches apart. He then places the boards along the lines. What type of lines do they represent?

2 in

2 in

Two straight lines are called **parallel** if they run side by side and never cross—no matter how far they extend in either direction. || is the symbol for parallel lines. In the drawing, $A \| B$ or, line A and line B are parallel.

A

B

To determine if two lines are parallel, follow these steps:

STEP 1. Make sure both lines are straight lines.

STEP 2. Check to see if the lines are an equal distance apart at two points.

If both steps 1 and 2 are true, the lines are parallel. Remember, parallel lines never meet, no matter how far they are extended in either direction.

PRACTICE 6

Write *yes* if the lines are parallel; *no* if they are not parallel.

1.

Answer ________

2.

Answer ________

3.

Answer ________

4.

Answer ________

5.

Answer ________

6. Give three examples where you might see parallel lines in your home.

7. Give three examples where you might see parallel lines in your community.

Draw a line parallel to the given line.

8.

9.

10.

Use the sample section of a city map to answer items 11–14.

11. Is 30th St parallel to 24th St?

Dodge St

Pacific St

30th St 24th St Radial St

12. Is Radial St parallel to Pacific St?

13. Is Dodge St parallel to Pacific St?

14. Is 24th St parallel to Dodge St?

To check your answers, turn to page 113.

Perpendicular Lines

The assembly directions tell Hakim to place the shelf perpendicular to the sides of the bookcase. What does perpendicular mean?

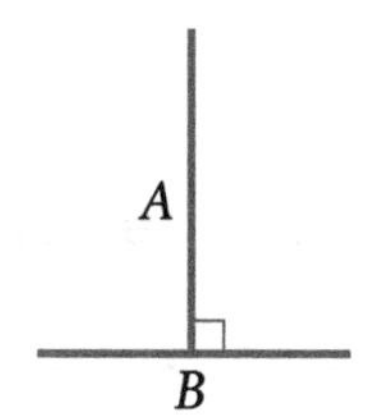

Unlike parallel lines, **intersecting lines** are straight lines which do cross each other. **Perpendicular lines** cross each other to form right angles (90° angles). The symbol for perpendicular lines is $\perp$. In the drawing at the right, $A \perp B$, or line A is perpendicular to line B.

Example: Determine if the shelf is perpendicular to the sides of the bookcase.

STEP 1. Measure the angle formed by the two straight lines.

STEP 2. If the lines intersect at a 90° angle, the two lines are perpendicular.

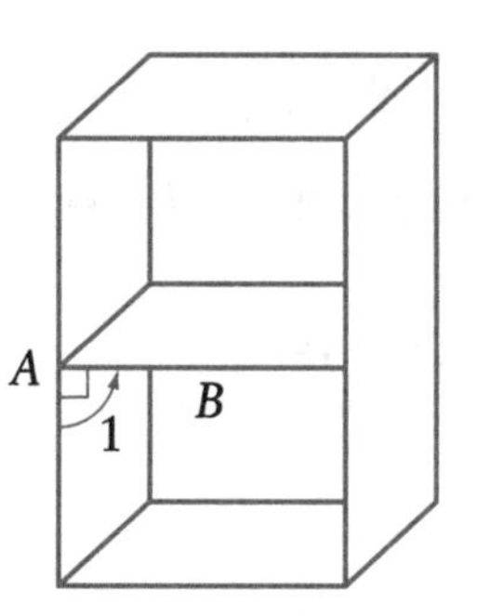

➠ Hakim finds that $A \perp B$, because $\angle 1$ is 90°.

PRACTICE 7

Write *yes* if the lines are $\perp$; *no* if they are not $\perp$.

1.

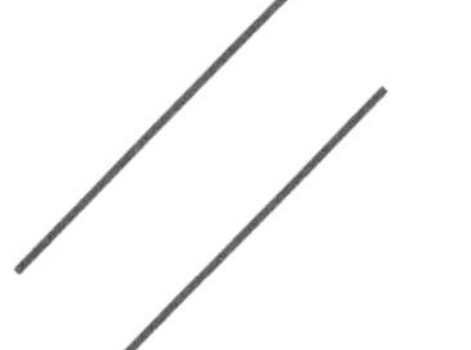

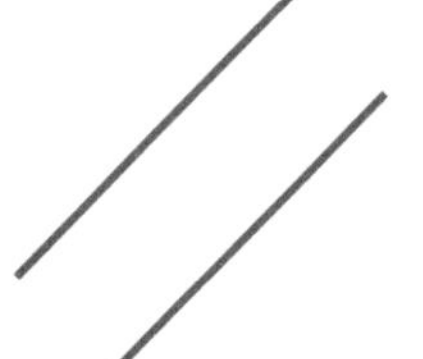

Answer ________

2.

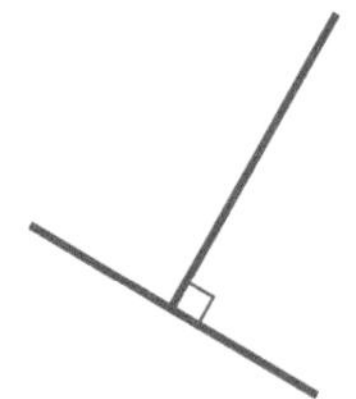

Answer ________

3.

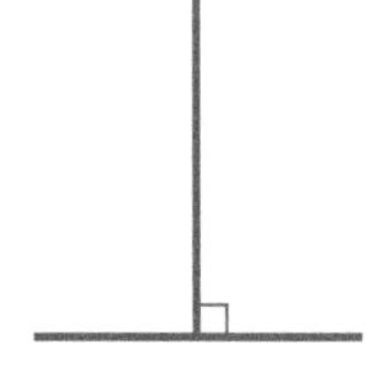

Answer ________

4.

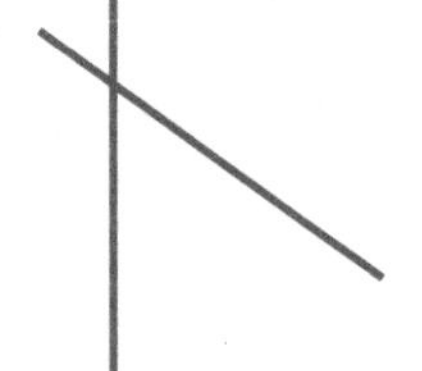

Answer ________

5.

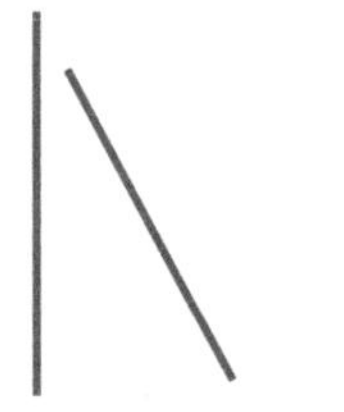

Answer ________

6.

Answer ________

7. Give three examples where you might see $\perp$ lines in your home.

8. Give three examples where you might see $\perp$ lines in your community.

Draw a line perpendicular to the given line.

9. 

10. 

11. 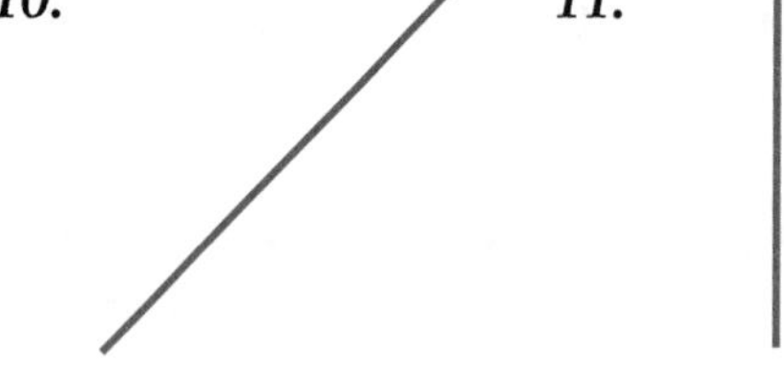

Use this section of the city map to answer items 12–17.

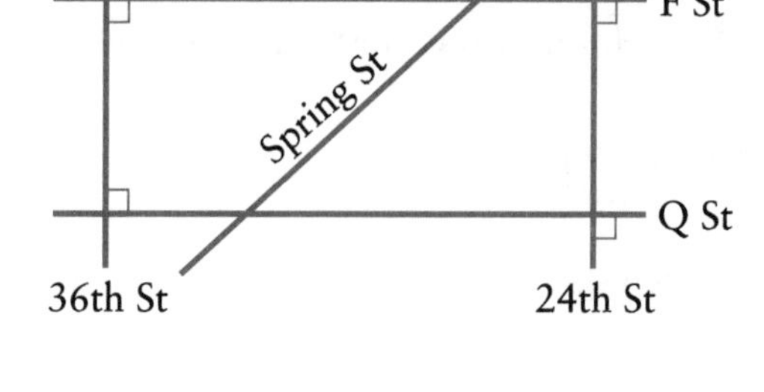

12. Is Q $\perp$ 36th?

13. Is Q $\perp$ F?

14. Is Q $\perp$ Spring?

15. Is F $\perp$ 24th?

16. Is F $\perp$ Spring?

17. Is F $\perp$ 36th?

To check your answers, turn to page 113.

Transversals and Vertical Angles

A tree falls across a street as in the illustration. From an aerial view, the street curbs appear as parallel lines. What kind of line does the tree make?

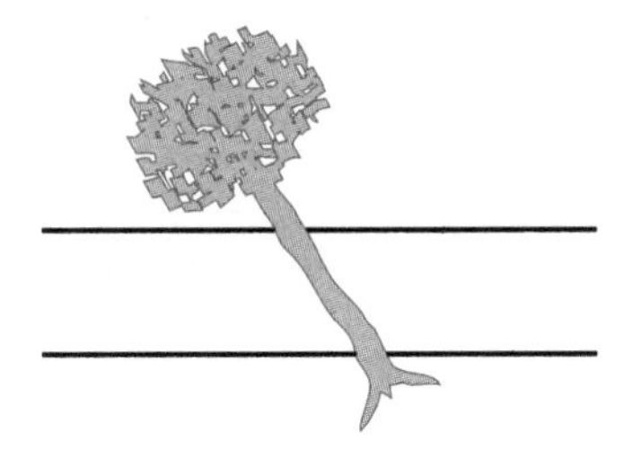

A **transversal** is a line that crosses parallel lines. The transversal forms the same angles where it crosses both of the parallel lines. In the illustration below, angles $\angle 1$ and $\angle 5$ are equal. These angles are called **corresponding angles** because they have the same relationship to the transversal and parallel lines. Other corresponding pairs of angles are $\angle 2$ and $\angle 6$, $\angle 3$ and $\angle 7$, and $\angle 4$ and $\angle 8$.

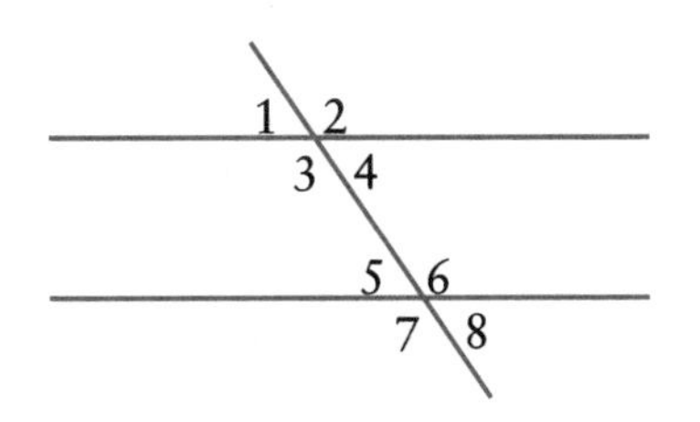

The angles across from each other are also equal to each other and are called **vertical angles.** In the illustration, $\angle 1$ and $\angle 4$ are vertical angles. Other pairs of vertical angles are $\angle 2$ and $\angle 3$, $\angle 5$ and $\angle 8$, and $\angle 6$ and $\angle 7$.

The angles on the same side of the transversal and parallel lines are supplementary. $\angle 1$ and $\angle 2$ are supplementary angles—they add up to 180°. Other supplementary pairs of angles are $\angle 3$ and $\angle 4$, $\angle 5$ and $\angle 6$, $\angle 7$ and $\angle 8$, $\angle 3$ and $\angle 5$, and $\angle 6$ and $\angle 8$.

If you know the number of degrees of one of the angles, you can find all of the other angles—they will either be equal to or supplementary to the known angle.

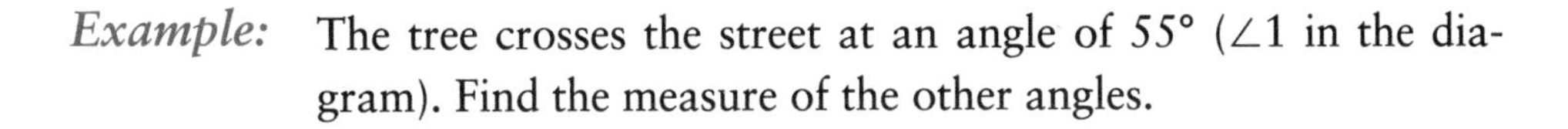

Example: The tree crosses the street at an angle of 55° ($\angle 1$ in the diagram). Find the measure of the other angles.

Step 1. $\angle 5$ corresponds to $\angle 1$, so it equals 55°.

Step 2. The vertical angle to $\angle 1$ is $\angle 4$; the vertical angle to $\angle 5$ is $\angle 8$, so they also equal 55°.

Step 3. Subtract 55° from 180° because $\angle 2$, $\angle 3$, $\angle 6$, and $\angle 7$ are supplementary to $\angle 1$. $180° - 55° = 125°$

➡ $\angle 1$, $\angle 4$, $\angle 5$, and $\angle 8$ are 55°, while $\angle 2$, $\angle 3$, $\angle 6$, and $\angle 7$ are 125°.

PRACTICE 8

Use the diagram to answer 1–10

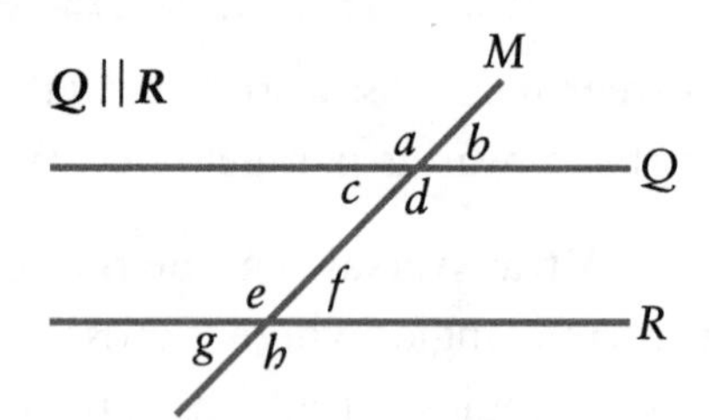

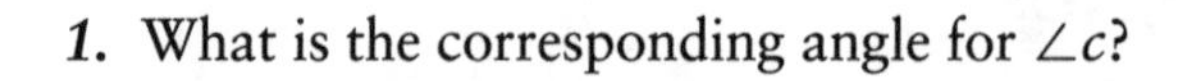

1. What is the corresponding angle for $\angle c$?

2. What is the corresponding angle for $\angle e$?

3. What is the corresponding angle for $\angle b$?

4. What kind of lines are Q and R?

5. What is the vertical angle for $\angle g$?

6. What is the vertical angle for $\angle d$?

7. What is the vertical angle for $\angle c$?

8. What angles have the same measure as $\angle f$?

9. What angles have the same measure as $\angle a$?

10. If $\angle e = 120°$, what is the measure of every other angle?

Find the measure of each angle for items 11 and 12.

11. X||Y 

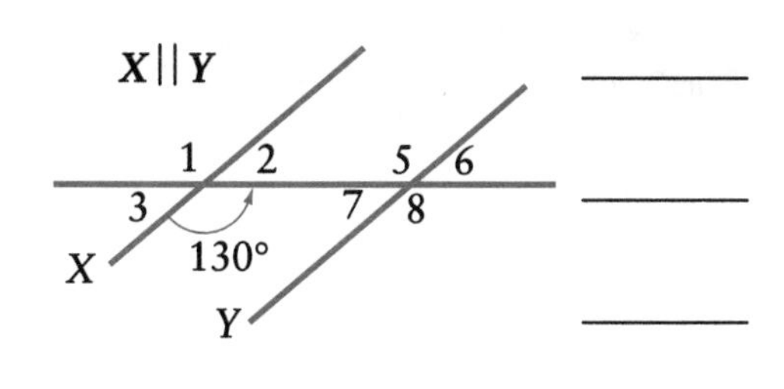

12. M||N

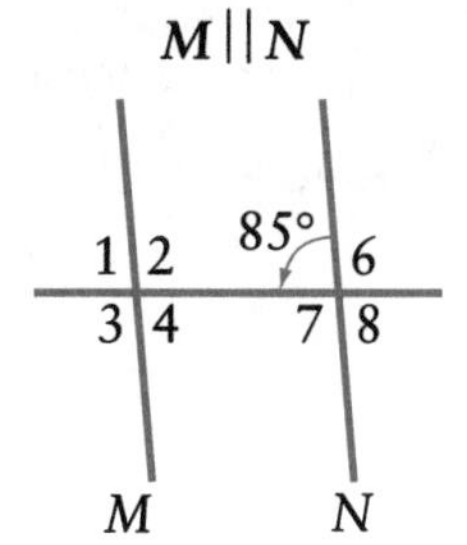

The two main wires for the cable company must be parallel to work efficiently. Connecting wires can cross at any angle. Use the diagram to answer items 13–18.

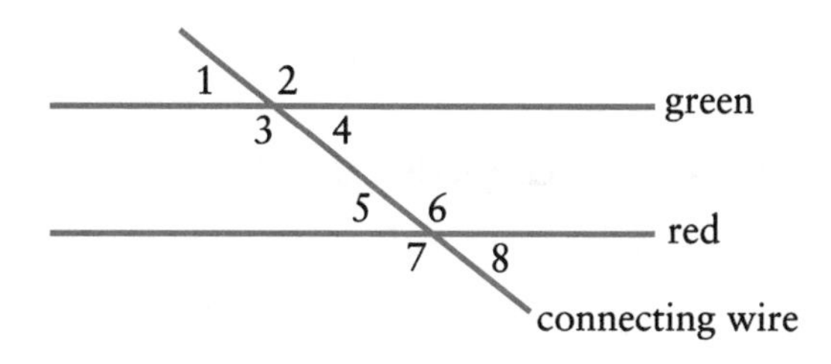

13. What angles have the same measure as $\angle 8$?

14. What angles have the same measure as $\angle 7$?

15. What is the corresponding angle of $\angle 2$?

16. What is the vertical angle of $\angle 1$?

17. If $\angle 4$ measures 60°, what is its supplement?

18. If $\angle 1$ measures 50°, what is the measure of the other angles?

To check your answers, turn to page 114.

ANGLES AND LINES REVIEW

The problems on pages 26 to 28 will help you find out if you need to review the angles and lines section of this book. When you finish, look at the chart on page 28 to see which pages you should review.

Label each angle.

1. 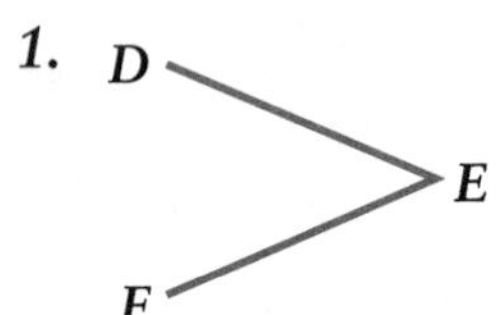

____

or

____

2. 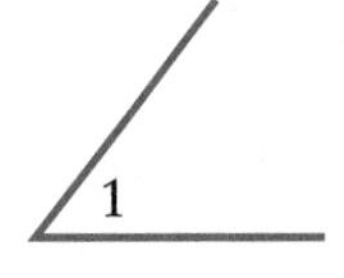

____

3. 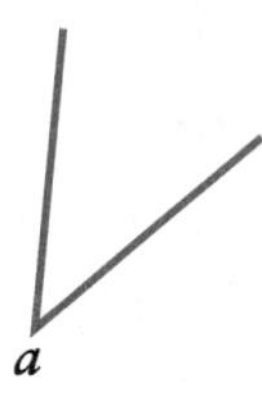

____

What kind of angle is each?

4. 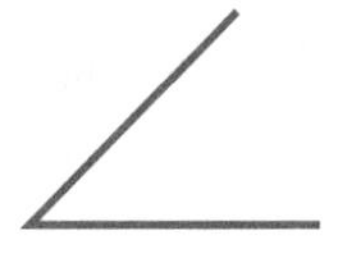

____

5. 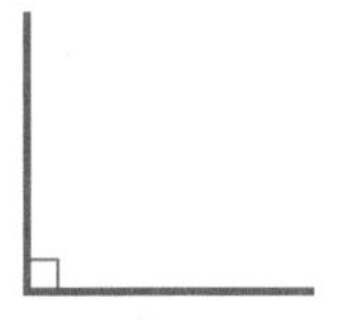

____

6. 

____

7. 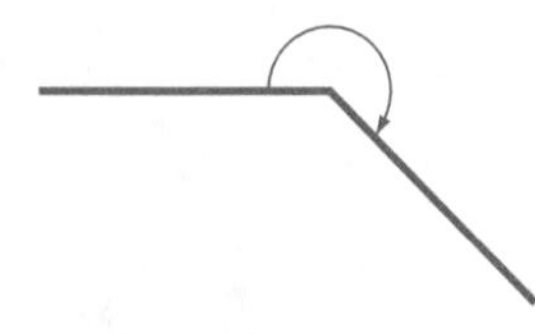

____

8. 

____

9. 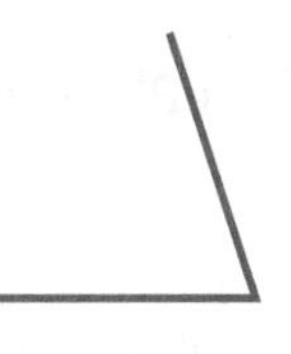

____

Find the complement of each angle in items 10–12.

10. 25° ________ *11.* 84° ________ *12.* 51° ________

Find the supplement of each angle in items 13–15.

13. 30° ________ *14.* 62° ________ *15.* 115° ________

Find the complement and supplement of each angle in items 16–18.

16. 36° ________ *17.* 74° ________ *18.* 58° ________

Match the two columns.

________ *19.* acute angle — *a.* <180° and >90°

________ *20.* reflex angle — *b.* 90°

________ *21.* parallel lines — *c.* <90°

________ *22.* obtuse angle — *d.* crosses parallel lines

________ *23.* vertical angles — *e.* 2 ∠s that total 180°

________ *24.* complementary angles — *f.* similar position on parallel lines

________ *25.* perpendicular lines — *g.* >180° and <360°

________ *26.* straight angle — *h.* lines that form a right angle

________ *27.* supplementary angles — *i.* equal to each other

________ *28.* corresponding angles — *j.* 180°

________ *29.* right angle — *k.* lines that never meet

________ *30.* transversal — *l.* 2 ∠s that total 90°

Use the diagram to answer problems 31–38.

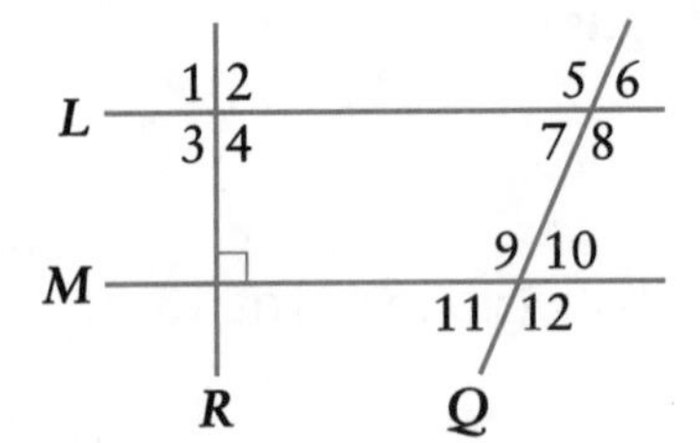

31. What kind of lines are L and M?

32. What kind of lines are M and R?

33. What kind of line is Q?

34. What is the vertical angle for $\angle 8$?

35. What is the corresponding angle for $\angle 11$?

36. What is the measure of $\angle 2$?

37. Which angles have the same measure as $\angle 8$?

38. If $\angle 6$ is 70°, what is $\angle 8$?

PROGRESS CHECK

Check your answers on page 114. Then return to the review pages for the problems you missed. Correct your answers before going on to the next unit.

If you missed problems	*Review pages*
1 to 3	7 to 9
4 to 9, 19 to 20, 22, 26, or 29	13 to 15
10 to 18, 24, or 27	15 to 18
21, 23, 25, 28, 30 to 38	19 to 25

UNIT 2

Triangles

Kinds of Triangles

This picture shows a drawing of Jack's house. The roof of his house forms a figure made up of three straight lines. What type of figure is this?

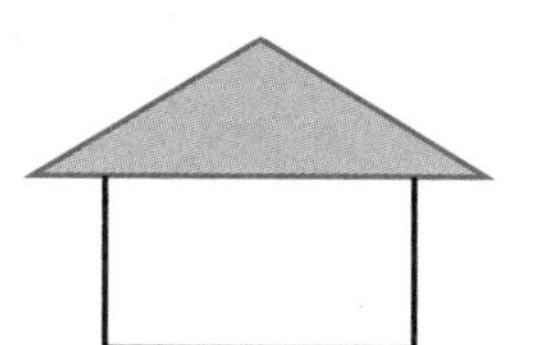

Triangles are figures formed by three straight lines. There are four kinds of triangles. An **equilateral triangle** has three sides equal in length and three equal angles. **Isosceles triangles** have two equal sides and two equal angles.

A triangle with one right (90°) angle is a **right triangle.** The fourth kind of triangle is a **scalene triangle.** Scalene triangles have no equal angles and no equal sides.

To show that a figure is a triangle, use the $\triangle$ symbol followed by the letters of the three points. $\triangle ABC$ stands for each triangle below.

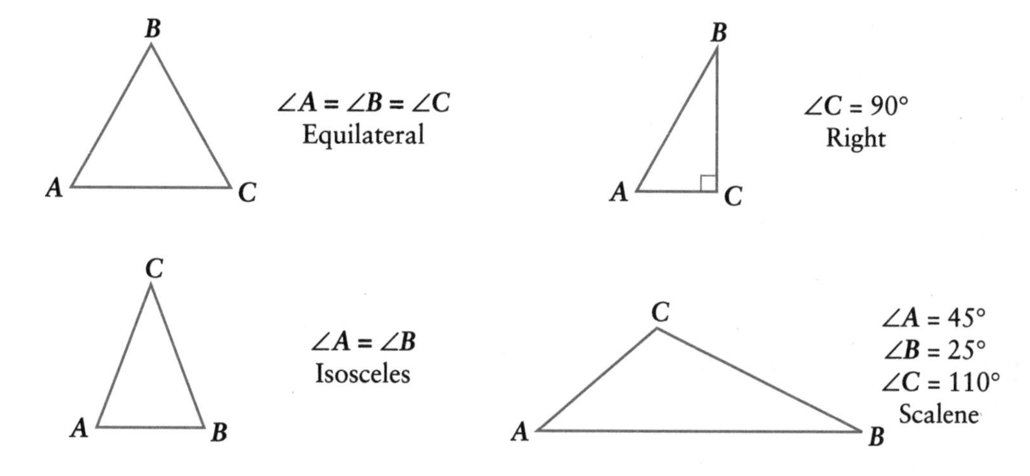

Example: Determine the kind of triangle that is formed by the roof of Jack's house.

[Hint: If angles are too small to measure, extend the lines before using the protractor.]

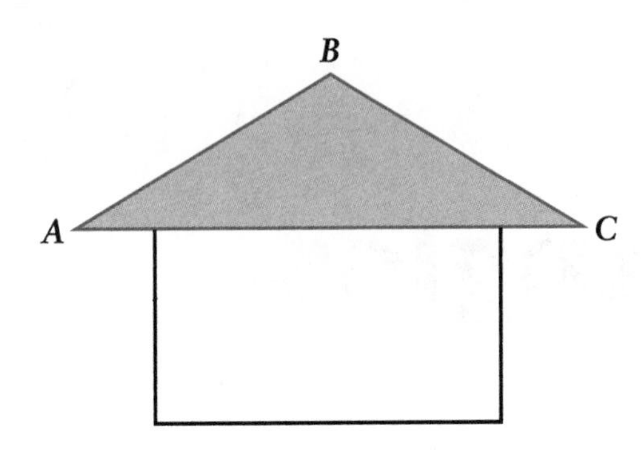

STEP 1. Measure the angles inside the triangle.

$\angle A$ is 30°
$\angle B$ is 120°
$\angle C$ is 30°

STEP 2. Find the kind of triangle using the table below.

Description	*Type of Triangle*
Three equal angles	Equilateral triangle
Two equal angles	Isosceles triangle
One right angle	Right triangle
No equal angles	Scalene triangle

➠ Jack's roof has two equal angles, so it is an isosceles triangle.

PRACTICE 9

Match the definition with the triangle.

________ **1.** equilateral triangle   *a.* two equal sides

________ **2.** right triangle   *b.* all sides equal

________ **3.** isosceles triangle   *c.* triangle with a 90° angle

________ **4.** scalene triangle   *d.* all sides different

Label each triangle, equilateral, right, isosceles, or scalene.

5.

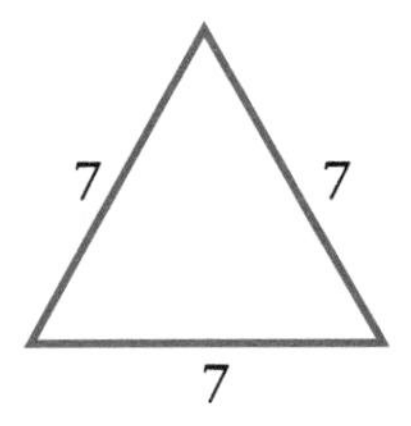

____

6.

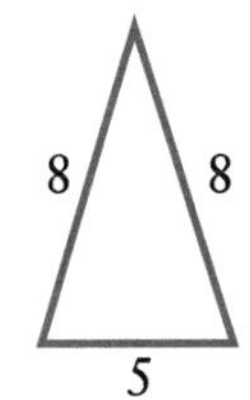

____

7.

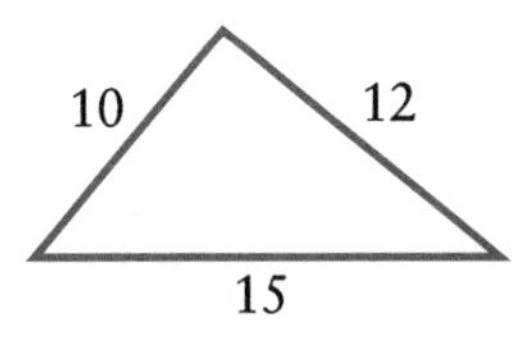

____

8.

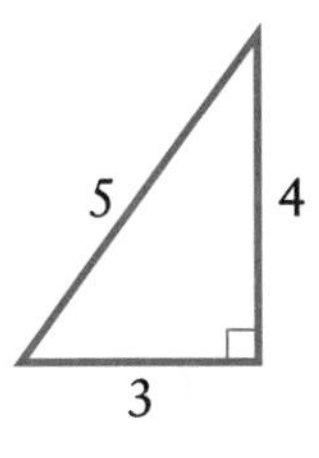

____

9.

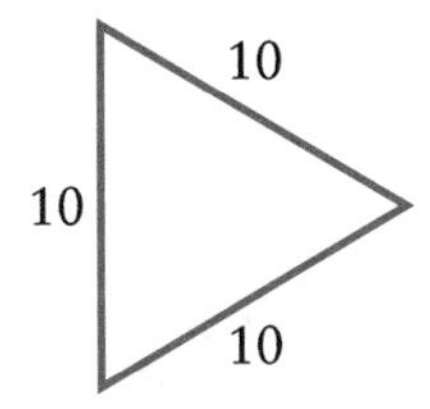

____

10.

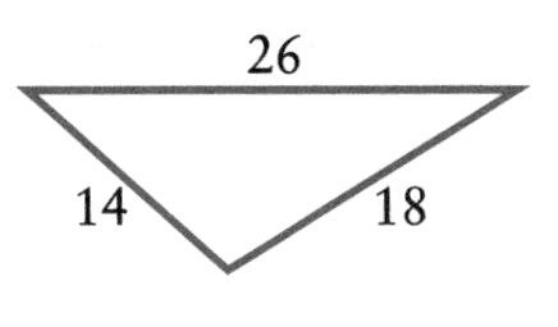

____

11.

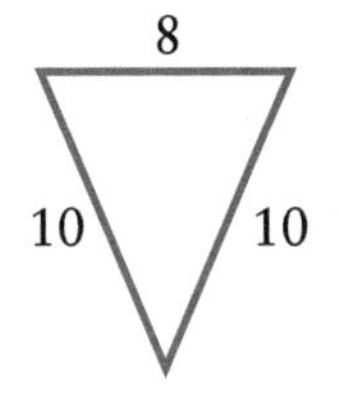

____

12.

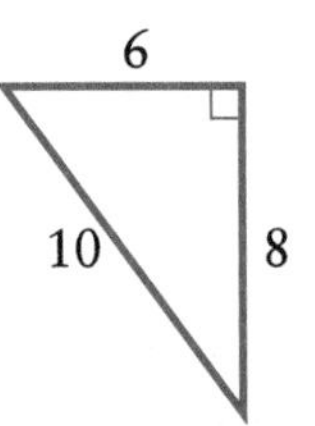

____

13.

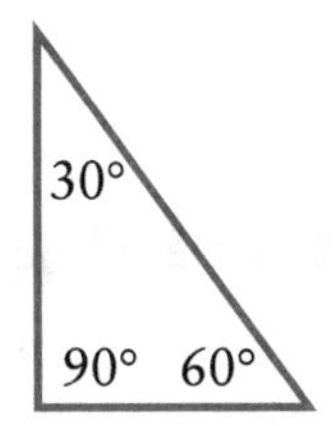

____

Measure the length of each side. Then label the triangle.

14.

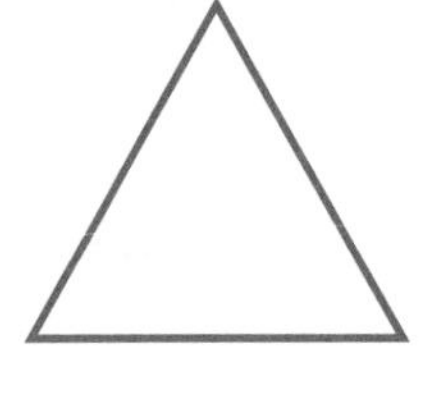

____

15.

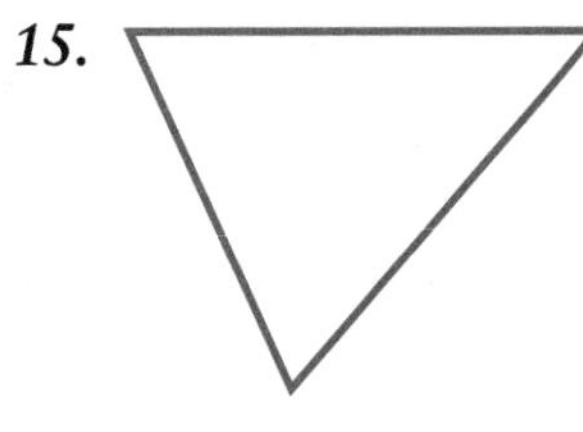

____

16. 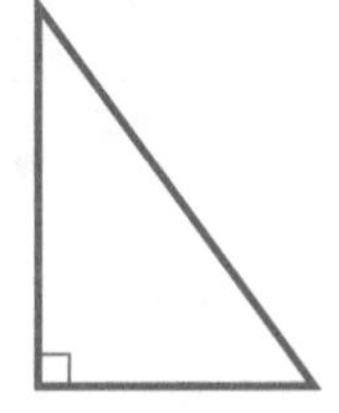

____

17. 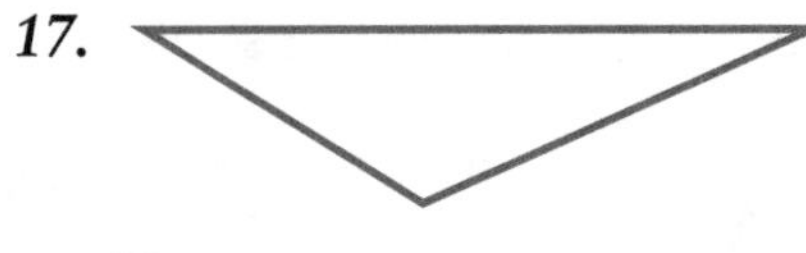

____

18. Jeri was making a collage with colored tissue paper. She decided all pieces should be triangular. If the two angles on each blue piece are 80°, what type triangle will each blue piece be?

19. The orange pieces have three different lengths for the sides. What type of triangle is an orange piece?

To check your answers, turn to page 115.

The Angles in a Triangle

Jason has cut the triangle at the right from a piece of cardboard. None of the sides of the triangle are equal and two of the angles measure 50° and 75°. What is the measure of the third angle?

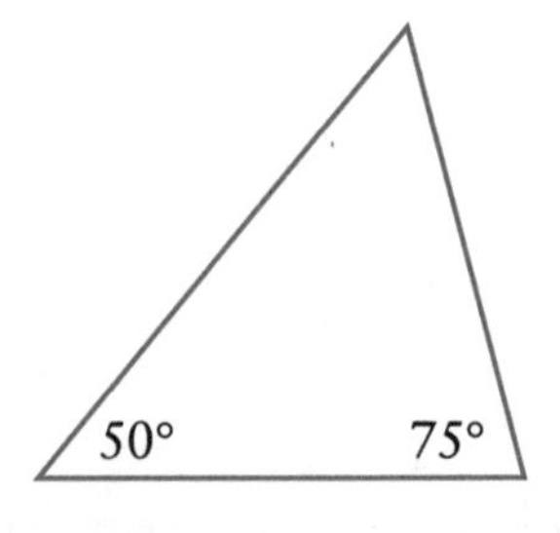

The three angles inside a triangle always add up to 180°. This is true for any type of triangle.

Example: Find the number of degrees in the third angle of the triangle.

STEP 1.	Add the number of degrees in the two known angles.	50° + 75° = 125°
STEP 2.	Subtract the result in step 1 from 180°.	180° − 125° = 55°

➠ The third angle in Jason's triangle is 55°.

PRACTICE 10

What is the measure of the unlabeled angle in each triangle?

1. 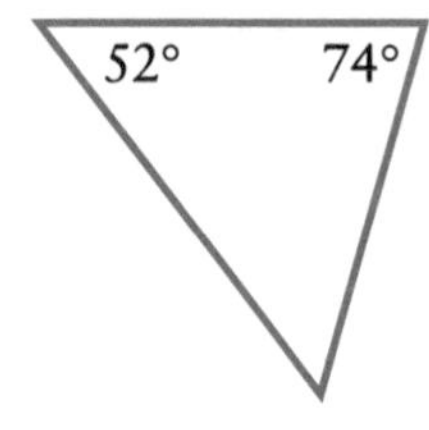

____

2. 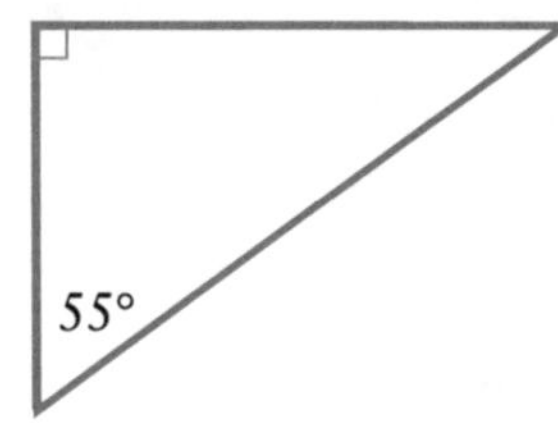

____

3. 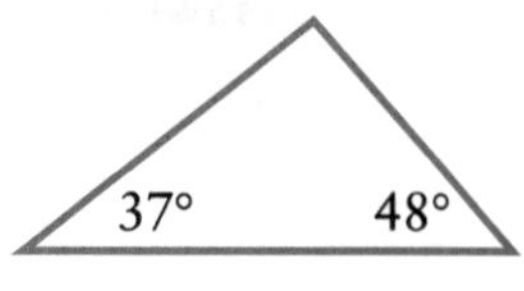

____

4. 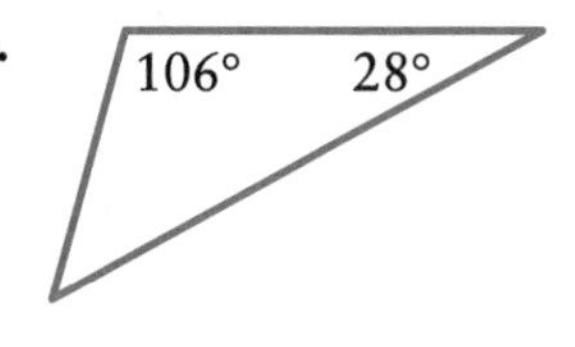

____

5. 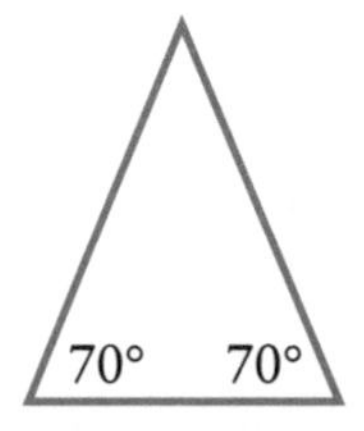

____

6. 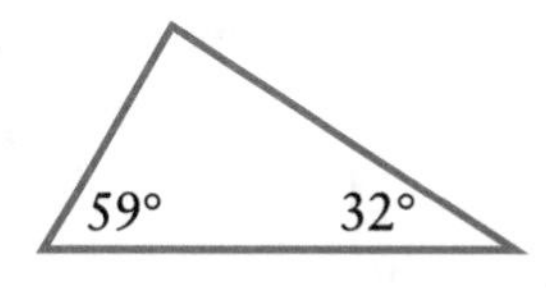

____

Solve the following.

7. $\triangle ABC$ is a right triangle. If $\angle C = 48°$, what is $\angle A$?

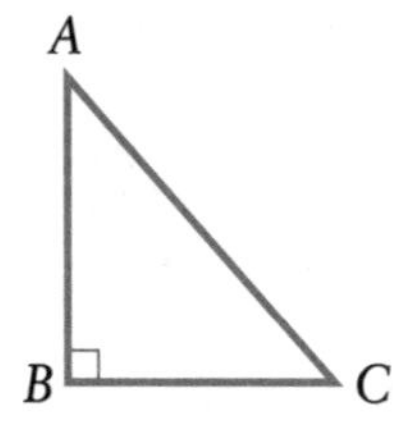

8. $\triangle PQR$ is an isosceles triangle. If $\angle R = 40°$, what is $\angle P$?

what is $\angle Q$?

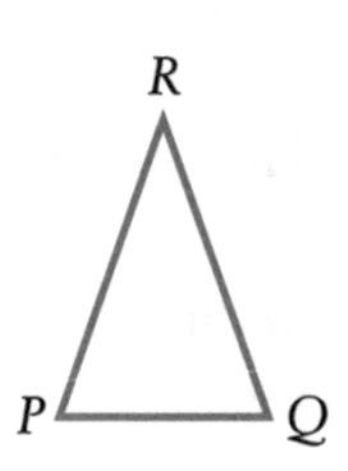

9. Jeremy was welding pieces of metal into a triangle. Two lengths were 18" and the third was 15". If the measure of each of the two equal angles is 62°, what is the measure of the third angle he is welding?

To check your answers, turn to page 115.

The Perimeter and Area of a Triangle

Drew's garden is in the shape of a triangle. A fence and cover are needed to protect it during the winter. What is the area of the triangle? What is the perimeter of the triangle?

The **perimeter** is the total measure of the sides of an enclosed figure. To find the perimeter (P) of a triangle, add the measure of the three sides, a, b, and c. The formula for the perimeter of a triangle is $P = a + b + c$.

The perimeter is measured in linear units. For example, if the sides are measured in feet, the perimeter is given as feet.

The **area** is the amount of surface space enclosed by the figure. To find the area (A) of a triangle, multiply $\frac{1}{2}$ times the base (b) times the height (h). Any side can be the base of a triangle. The height is perpendicular to the base. It reaches the point where the other two sides meet.

The formula for the area of a triangle is $A = \frac{1}{2}bh$.

The area is measured in square units. For example, if the sides are measured in inches, the area is given in square inches.

Example: Find the amount of fencing needed to enclose the garden shown in the diagram.

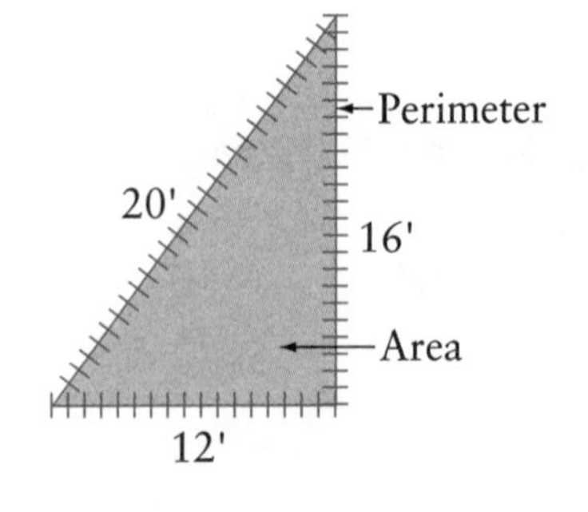

STEP 1. Find the measure of the sides of the triangle. — 12 ft, 16 ft, and 20 ft

STEP 2. Add the measures of the sides. — $P = 12 + 16 + 20$

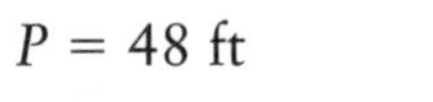

$P = 48$ ft

➠ The perimeter of the garden is 48 feet, which is the amount of fencing needed to enclose it.

Example: Find the area of the cover needed to protect the garden.

STEP 1. Find the height and base of the triangle. — height = 16 ft, base = 12 ft

STEP 2. Multiply $\frac{1}{2}$ times the height times the base.

$$A = \frac{1}{2}bh$$

$$A = \frac{1}{2} \times 16 \times 12$$

$$A = 96 \text{ sq ft}$$

Because this triangle is a right triangle, the height is one of the sides of the triangle.

➠ A 96-sq-ft cover is needed to protect the garden.

PRACTICE 11

A. Find the area and perimeter for each triangle.

1.

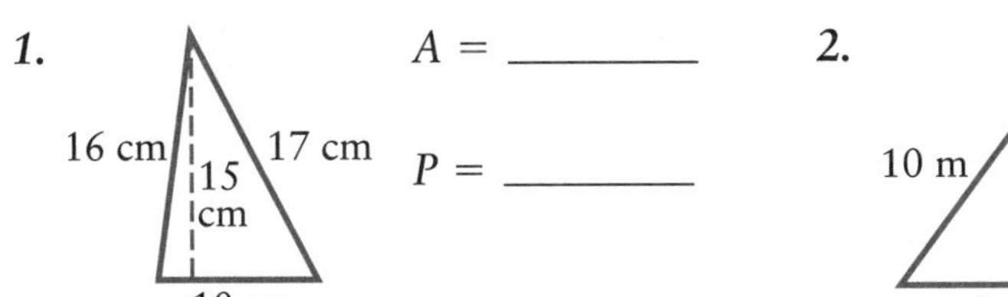

$A =$ ________

$P =$ ________

2.

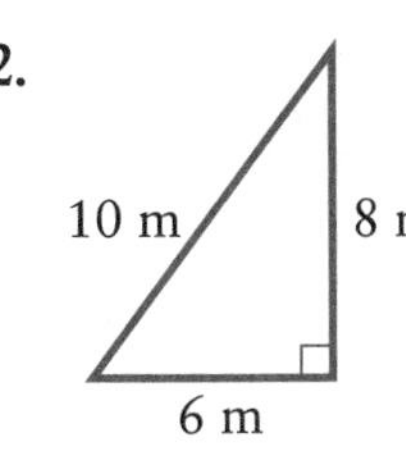

$A =$ ________

$P =$ ________

3.

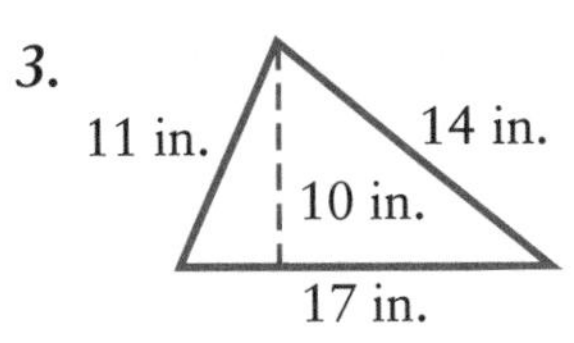

$A =$ ________

$P =$ ________

4.

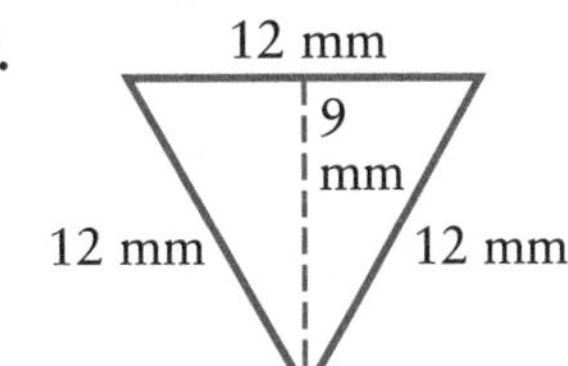

$A =$ ________

$P =$ ________

Edward measured the sides of the triangle as 2 feet, 2 feet, and 27 inches. The height was $1\frac{1}{2}$ feet. What is the perimeter of the triangle? What is the area?

All measurements must be in similar units before you can find the perimeter and area of a triangle.

> *Remember*
> 1 ft = 12 in
> 1 yd = 36 in
> 1 yd = 3 ft

Example: Find the perimeter of Edward's triangle.

STEP 1. Convert to similar units. $2' = 2 \times 12 = 24$ in

STEP 2. Substitute numbers in formula. $P = 24 + 24 + 27$

STEP 3. Complete the problem. $P = 75$ inches

➠ The perimeter is 75 inches.

Example: Find the area of Edward's triangle.

STEP 1. Convert to similar units. $1\frac{1}{2}\text{ ft} = 1\frac{1}{2} \times 12 = 18$ in

STEP 2. Substitute numbers in formula. $A = \frac{1}{2} \times 27 \times 18$

STEP 3. Complete the problem. $A = 243$ sq in

➠ The area is 243 sq in.

B. Find the area and perimeter of each triangle.

5. 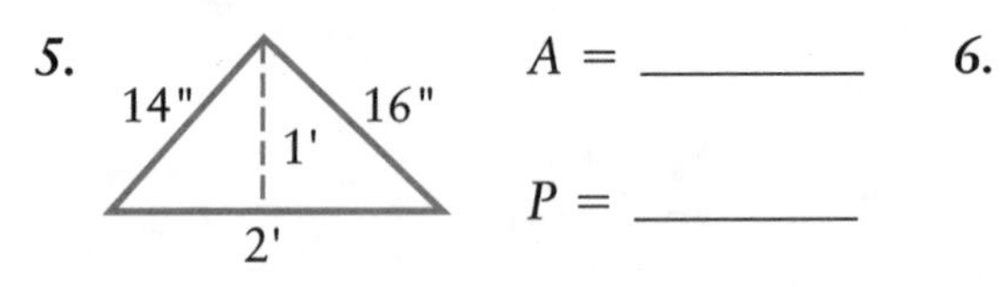

$A =$ ________ $P =$ ________

6. 3' 3' 30" 24"

$A =$ ________ $P =$ ________

7. 12' 14' 10' 6 yd

$A =$ ________ $P =$ ________

8. 2' 27 in. 18" 1 yd

$A =$ ________ $P =$ ________

To check your answers, turn to page 116.

Similar Triangles

Ken is making a sign for a store. He has cut a piece of glass into a triangle. He thinks his design would look best if he glued a smaller triangle with the same shape on top of the first.

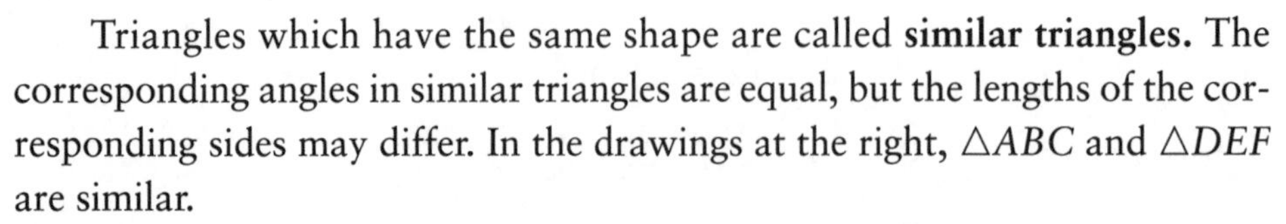

Triangles which have the same shape are called **similar triangles.** The corresponding angles in similar triangles are equal, but the lengths of the corresponding sides may differ. In the drawings at the right, $\triangle ABC$ and $\triangle DEF$ are similar.

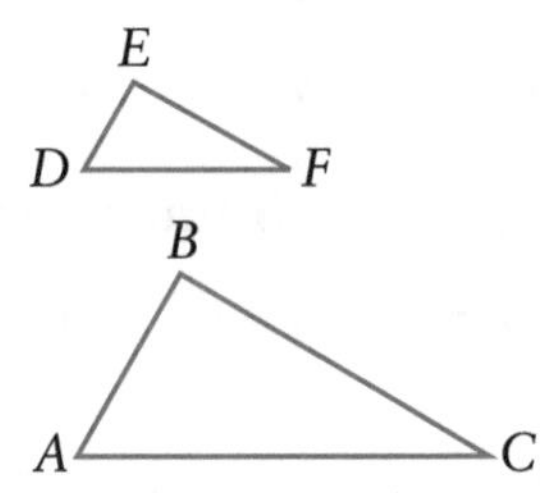

Because the corresponding angles in similar triangles are equal, the lengths have the same proportions. In the similar triangles in the diagram:

$$\frac{AB}{DE} = \frac{BC}{EF} \qquad \frac{AC}{DF} = \frac{BC}{EF}$$

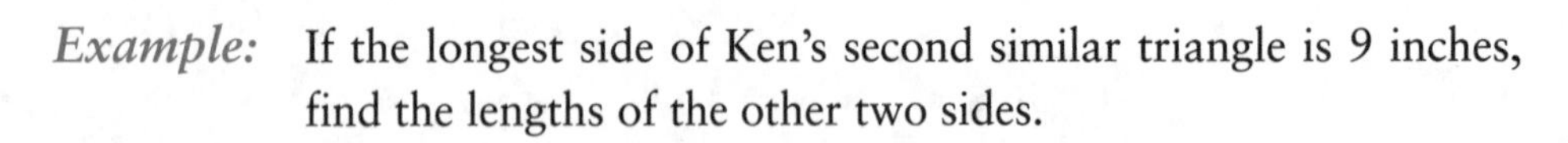

Example: If the longest side of Ken's second similar triangle is 9 inches, find the lengths of the other two sides.

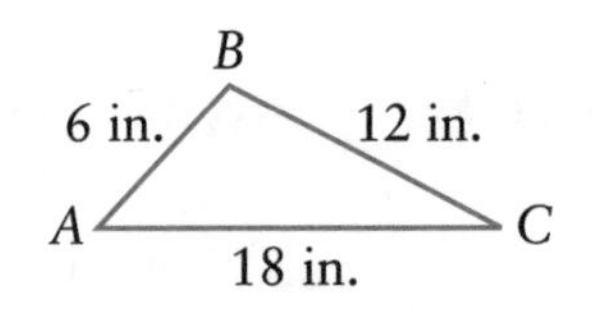

E ? ? D F 9 in.

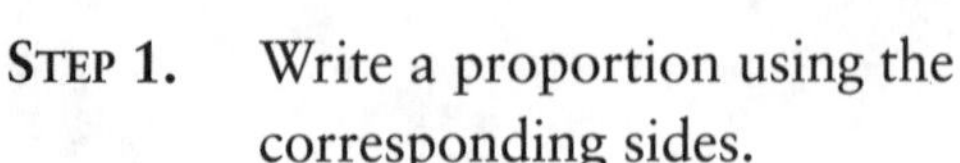

STEP 1. Write a proportion using the corresponding sides. $\frac{18}{9} = \frac{12}{x}$ $\frac{18}{9} = \frac{6}{y}$

STEP 2. Simplify. $2 = \frac{12}{x}$ $2 = \frac{6}{y}$

STEP 3. Solve the proportion. $x = 6$ $y = 3$

➡ The lengths of the other two sides must be 6 in and 3 in.

PRACTICE 12

Find the length of the unknown side in the similar triangles.

1.

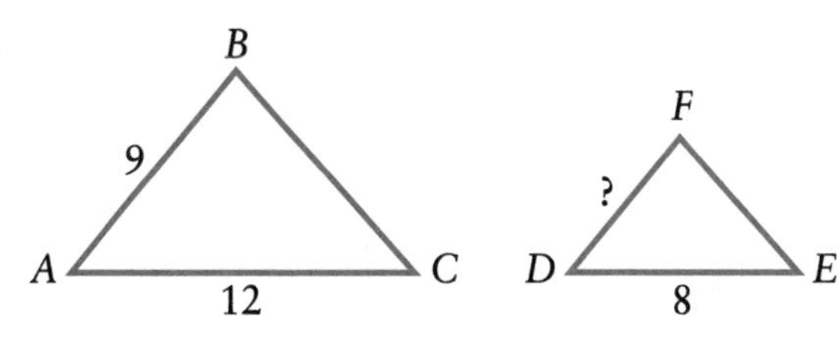

$\overline{DF}$= ________

2.

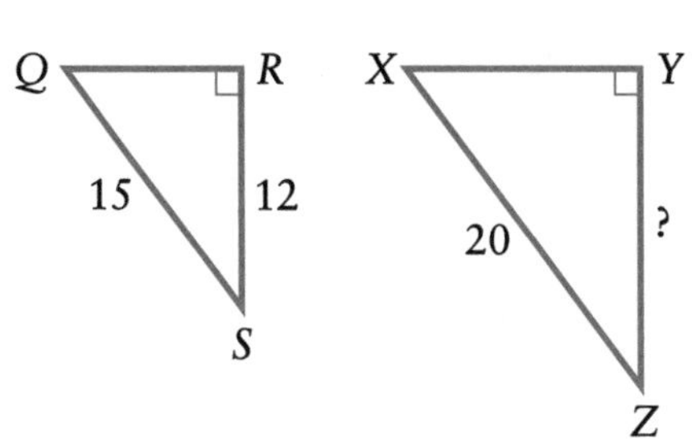

$\overline{YZ}$ = ________

3.

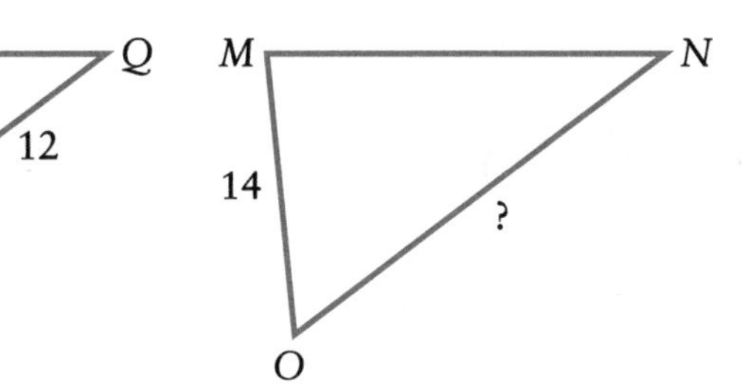

$\overline{NO}$ = ________

4.

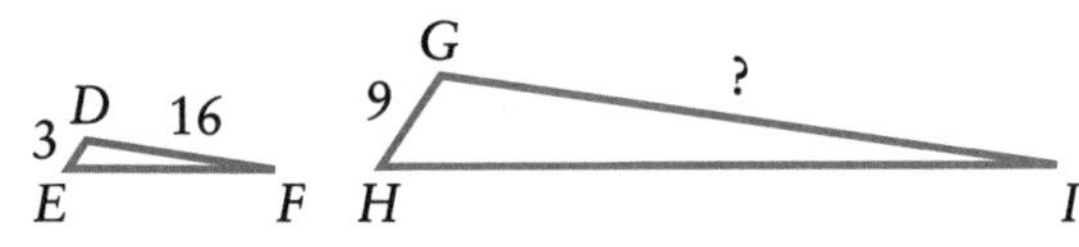

$\overline{GI}$ = ________

Find the lengths of the missing sides.

5.

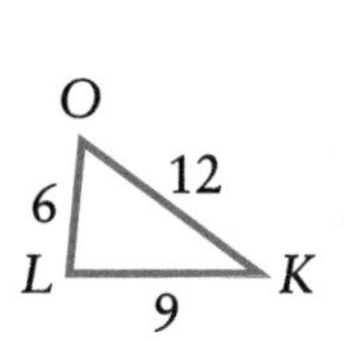

$\overline{RS}$ = ________

$\overline{RT}$ = ________

6.

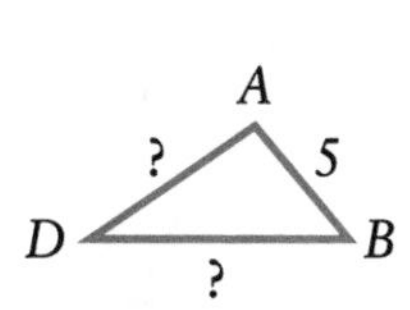

$\overline{AD}$ = ________

$\overline{DB}$ = ________

7.

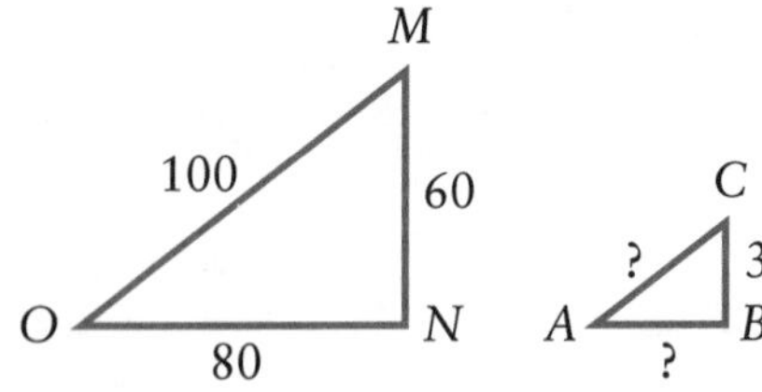

$\overline{AC}$ = ________

$\overline{AB}$ = ________

8. A city garden is triangular in shape with dimensions of 120 yards by 85 yards by 100 yards. Susan wants to make a model of the garden. If the longest side she can use is 24 inches, what will the other dimensions of the model be?

To check your answers, turn to page 116.

Equilateral Triangles

Jorge nails two boards of equal length together at a 60° angle. Now he wants to measure a third board to nail to the other two boards. What type of triangle will he make?

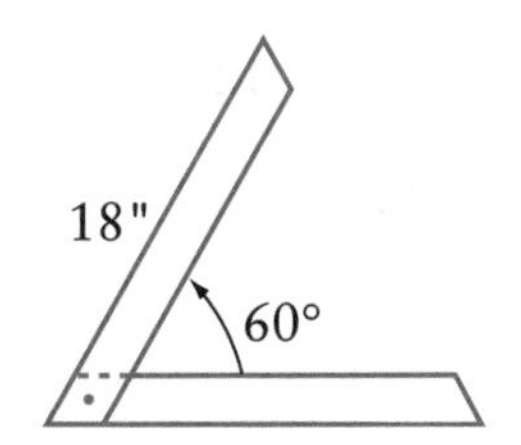

If two sides of a triangle are equal and form a 60° angle, the triangle must be an **equilateral triangle.** The third side will be the same length as the other sides, and all angles will measure 60°.

Since every side of one equilateral triangle is the same length, once you know the length of one side, you automatically know the lengths of the other sides. Therefore, the perimeter of an equilateral triangle is three times the length of any one of its sides.

Example: If the first two boards are 18 inches long and the angle formed by them is 60°, find the length of the third board.

STEP 1. Measure the length of either of the two boards. — Length = 18 inches

STEP 2. Find the angle formed by the two boards. — $\angle ABC = 60°$

STEP 3. If two sides are equal and form a 60° angle, the third side must be equal to the other two sides, forming an equilateral triangle. — AC = 18 inches

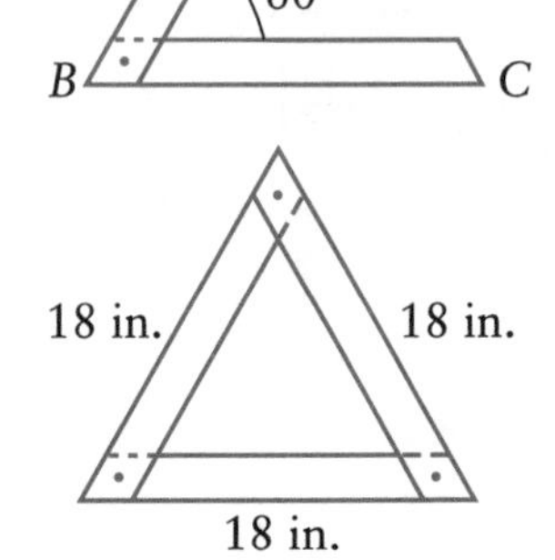

➡ Jorge must cut an 18-inch length of board. The triangle will be equilateral.

PRACTICE 13

A. Answer.

1. Each angle in an equilateral triangle is ______°.

2. Each side in an equilateral triangle has the ____________ measure.

3. The sum of the measures of the angles in an equilateral triangle is ______°.

Find the area and perimeter of each equilateral triangle.

4. 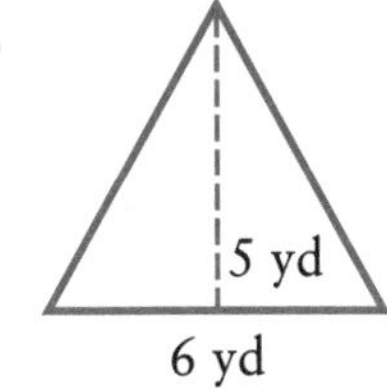

$A =$ ________

$P =$ ________

5. 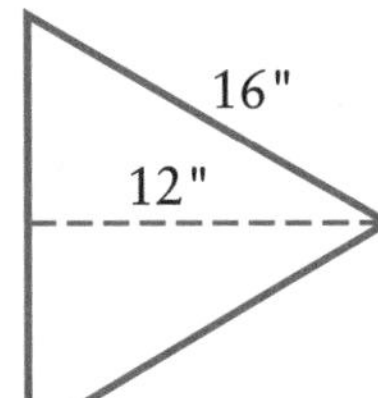

$A =$ ________

$P =$ ________

6. 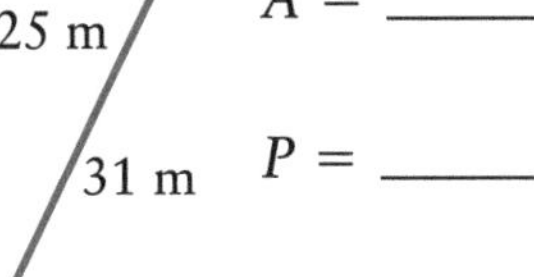

$A =$ ________

$P =$ ________

7. 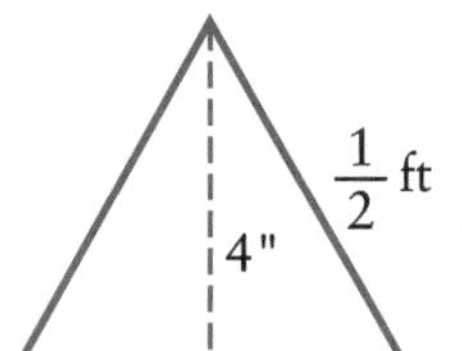

$A =$ ________

$P =$ ________

If the perimeter of an equilateral triangle is given, you can find the length of one side. Use the formula $P = 3s$, where s equals the length of one side.

Example: If the perimeter of an equilateral triangle is 36 inches, what is the length of one side?

STEP 1.	Write the formula.	$P = 3s$
STEP 2.	Substitute.	$36 = 3s$
STEP 3.	Solve the equation.	$\frac{36}{3} = \frac{3s}{3}$
		$12 = s$

➡ The length of one side is 12 inches.

B. Find the length of one side.

8.

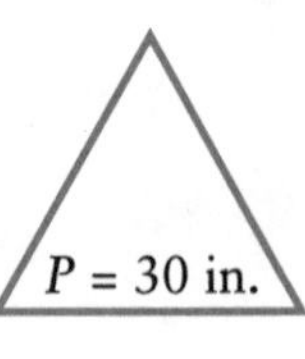

$s =$ ________

9.

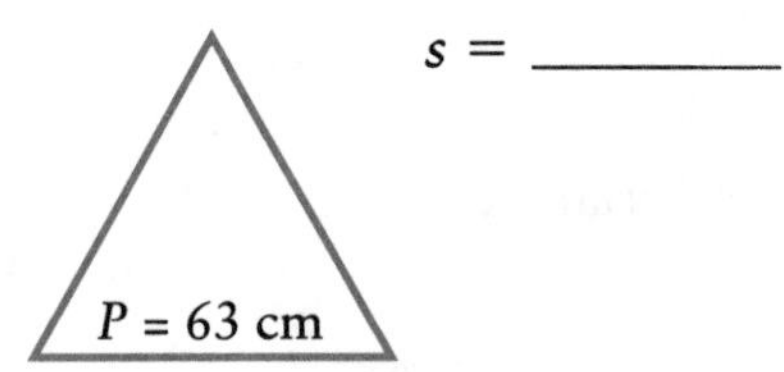

$s =$ ________

10.

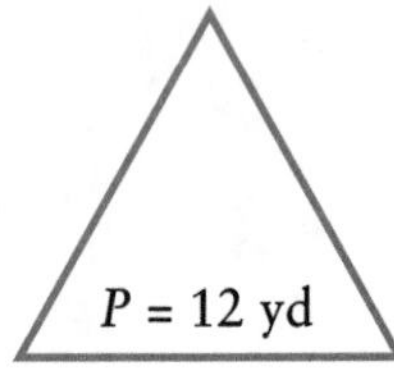

$s =$ ________

11.

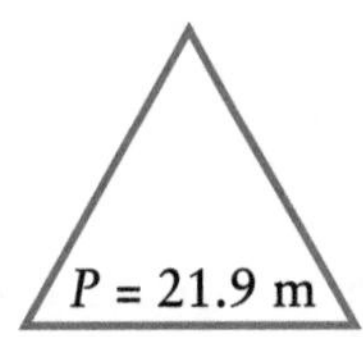

$s =$ ________

12.

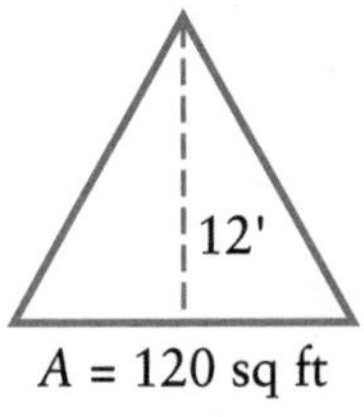

$s =$ ________

13.

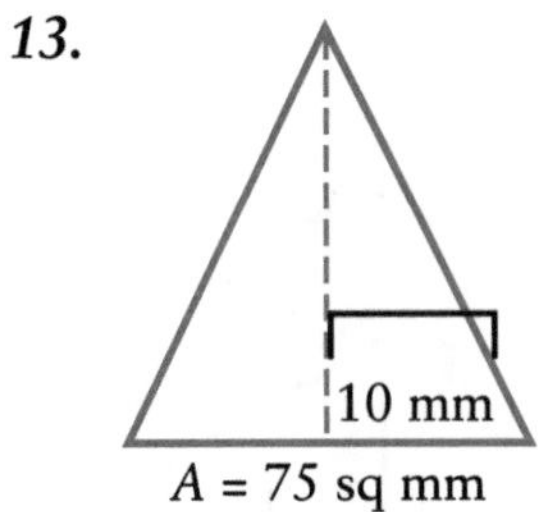

$s =$ ________

14.

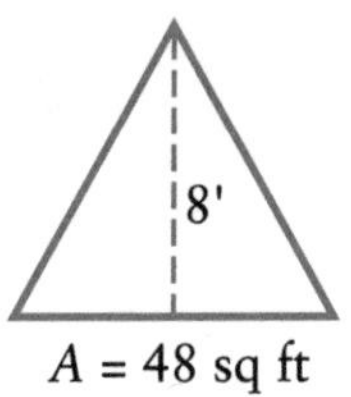

$s =$ ________

15. Kym wanted to put trim around the edges of a cloth shaped like an equilateral triangle. If she has 75 inches of trim, what is the maximum length of a side in the triangle?

16. The area of an equilateral triangle is 80 square inches. If the height of the triangle is 10 inches, what is the length of one side?

To check your answers, turn to page 117.

Isosceles Triangles

The roof of Mikko's house forms a 130° angle at the peak. The sides of the roof are equal in length and peak at the center of the house. Find the measures of the remaining angles.

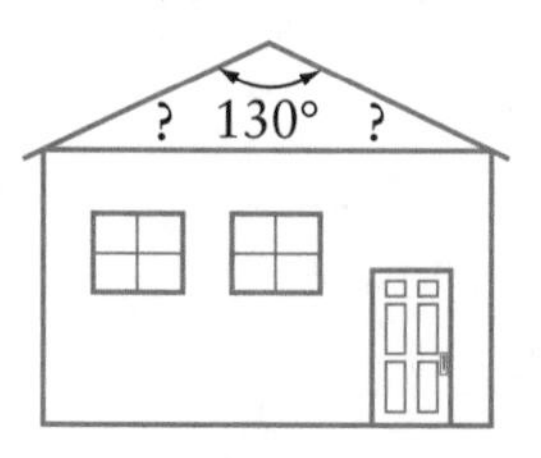

In an isosceles triangle, two of the sides are equal in length. The angles opposite the equal sides are equal as well. The equal angles are called the **base angles.** The third angle is known as the **vertex angle.**

Once you know one of the angles in an isosceles triangle, you can find the other two angles. If you know the vertex angle, each of the base angles can be found by subtracting the vertex angle from 180° and dividing the result by 2.

If you know one of the base angles, multiply by 2 and subtract the result from 180°. The answer is the vertex angle.

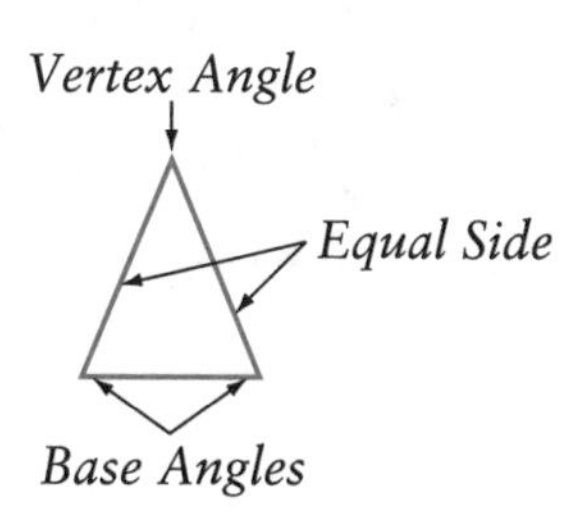

Example: Determine the base angles formed where the sides of Mikko's roof meet the rest of the house if the vertex angle is 130°.

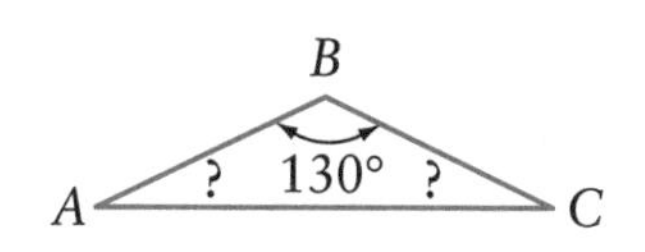

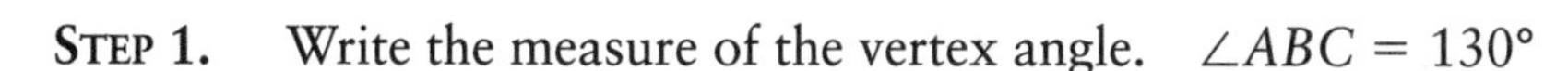

STEP 1. Write the measure of the vertex angle. $\angle ABC = 130°$

STEP 2. Subtract the measurement from 180°. $180° - 130° = 50°$

STEP 3. Divide the result by 2. $50 \div 2 = 25°$

STEP 4. The result is the base angle. $\angle BAC = 25°$ and $\angle BCA = 25°$

➡ The base angles of Mikko's roof are 25°.

PRACTICE 14

Write *yes* if the triangle is isosceles; *no* if it is not.

1.

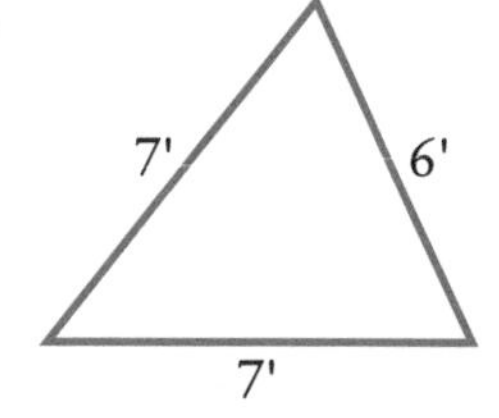

____

2.

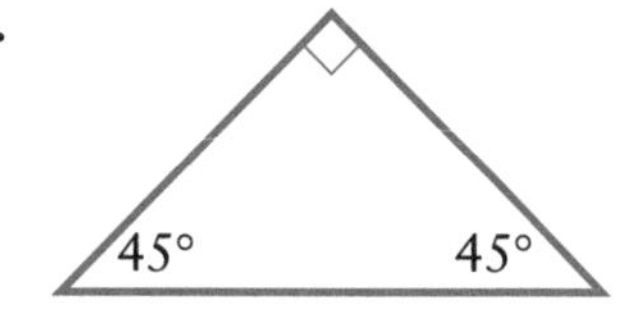

____

3.

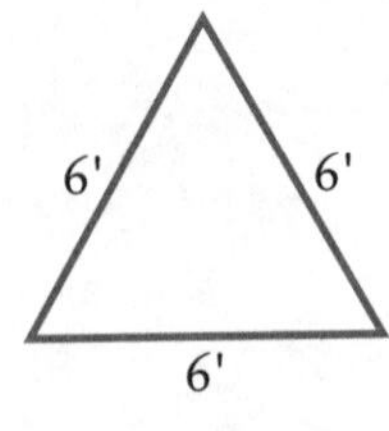

4.

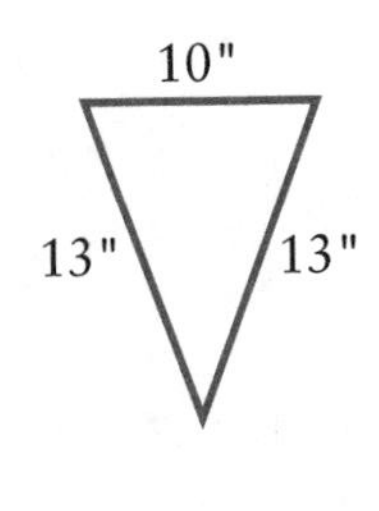

5.

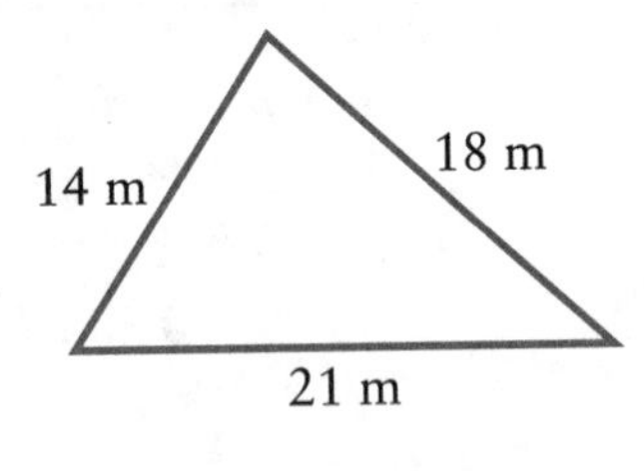

6.

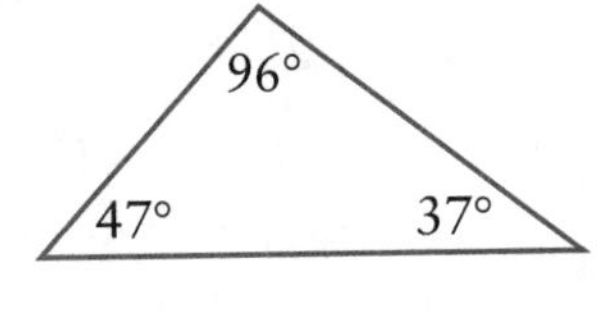

Find the measure of the base angles.

7.

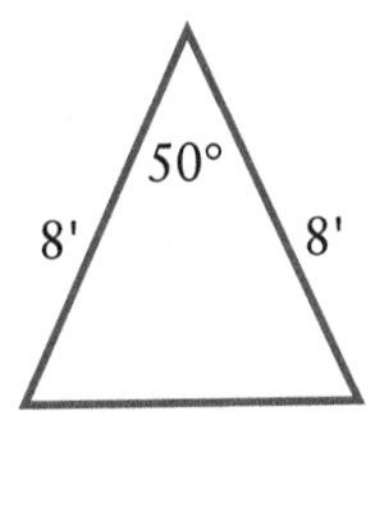

8.

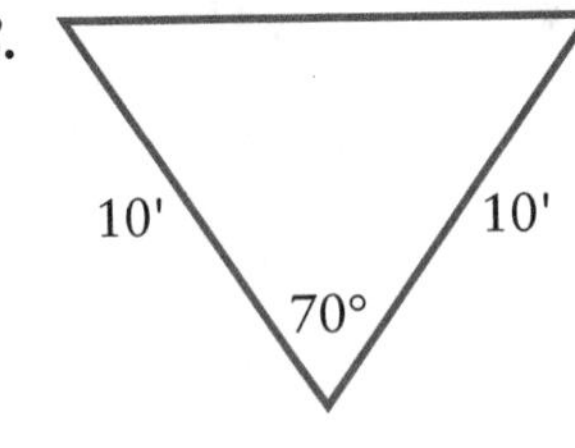

9.

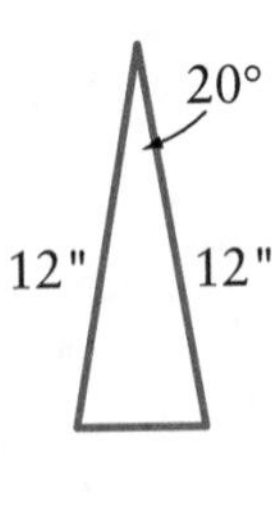

10.

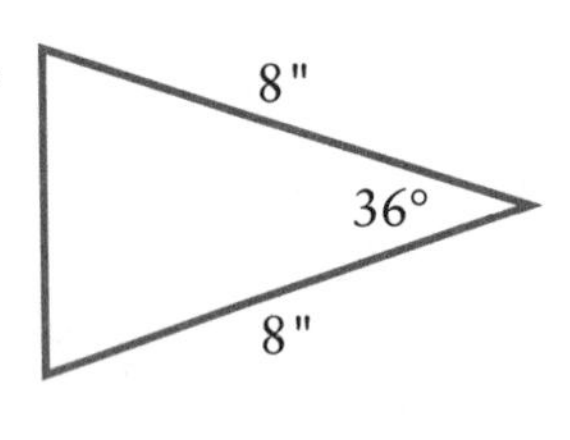

Find the measure of the vertex angle.

11.

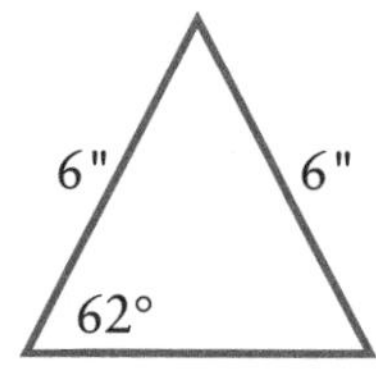

______

12.

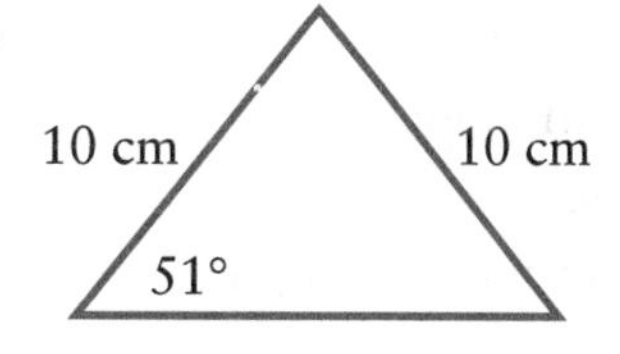

______

13.

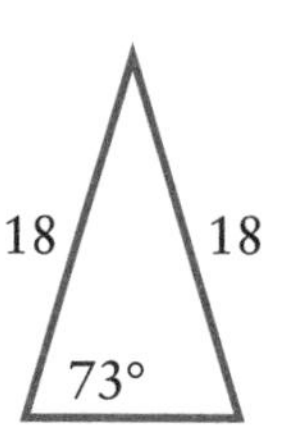

______

14.

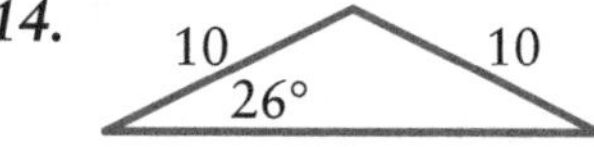

______

15. The top piece of glass in a stained glass window looks like an isosceles triangle. Sue measures the vertex angle at 68°. What is the measure of the base angles?

To check your answers, turn to page 118.

Right Triangles

Ben's driveway is perpendicular to the road. His walkway runs from the road to the end of the driveway as in the illustration. What type of triangle is formed?

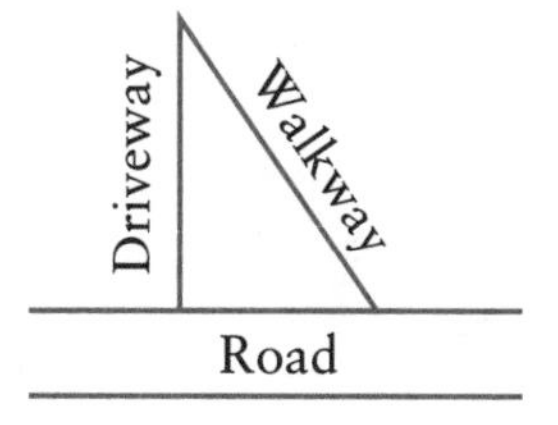

Perpendicular lines meet at a 90° angle. The triangle formed by the road, the driveway, and the walkway is a **right triangle**—a triangle with one 90° angle. The side opposite the right angle is called the **hypotenuse.** The other two sides are called **legs.** In the figure to the right, BC is the hypotenuse; AB and AC are the legs.

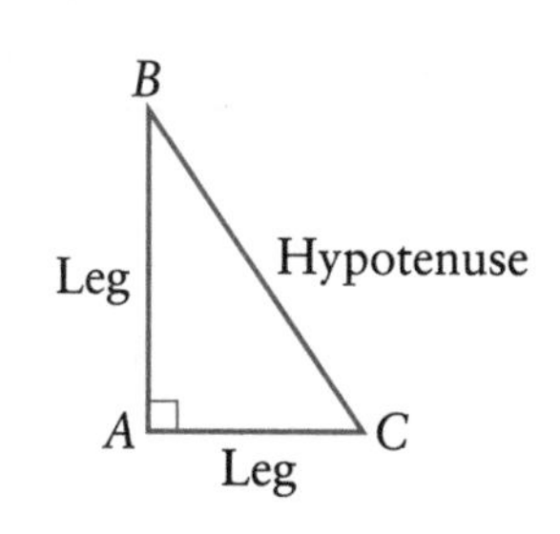

If you know one of the other angles in a right triangle, you can find the third angle. Since the three angles of a triangle add up to 180°, and one of the angles in a right triangle is always 90°, the other two angles must add up to 90°:

$$\begin{array}{rl} 180^\circ & \text{(sum of all angles)} \\ -\ \ 90^\circ & \text{(right angle)} \\ \hline 90^\circ & \text{(sum of other two angles)} \end{array}$$

Example: Ben's driveway and walkway meet at a 40° angle. Find the angle formed by the walkway and the road without measuring the unknown angle.

STEP 1. Measure one of the smaller angles. $\angle ABC = 40^\circ$

STEP 2. Subtract the answer to step 1 from 90°. $90^\circ - \angle ABC = \angle BCA$
$90^\circ - 40^\circ = 50^\circ$

➡ The angle formed by the walkway and the road is 50°.

PRACTICE 15

Fill in the blanks.

1. A right angle is ______°.

2. The side opposite the right angle in a right triangle is the __________.

3. The sum of the measures of the three angles in a right triangle is always ______°.

Write *yes* if the diagram is a right triangle; *no* if it is not.

4. 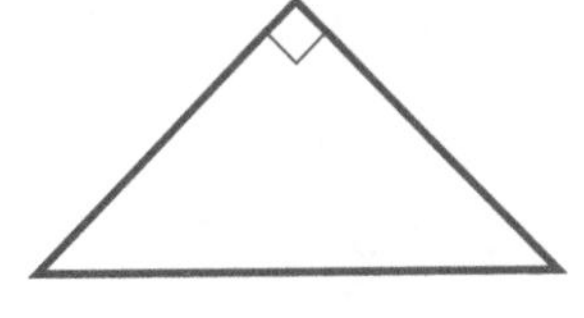

______

5. 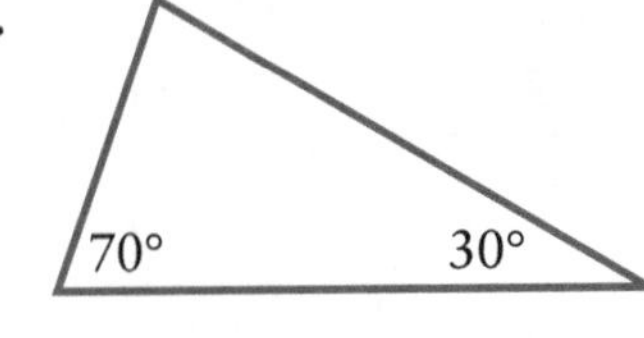

______

6.

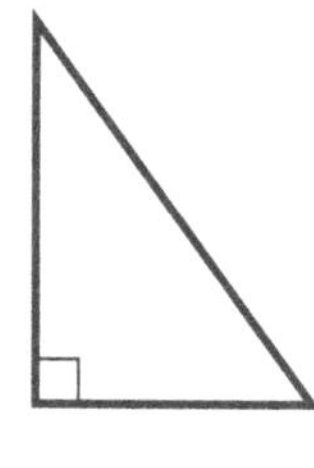

____

7.

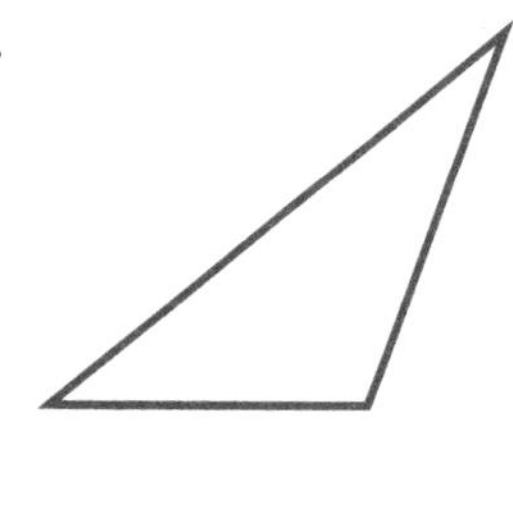

____

Find the measure of the third angle.

8.

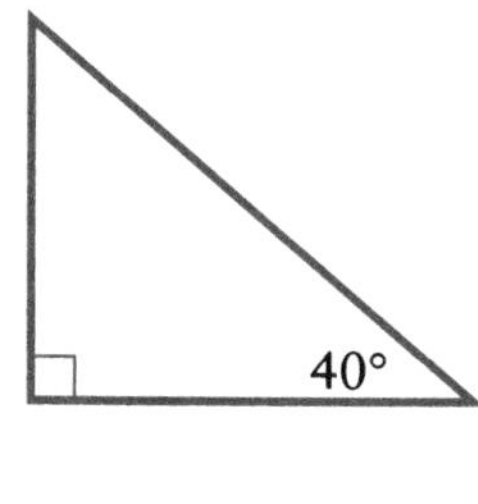

____

9.

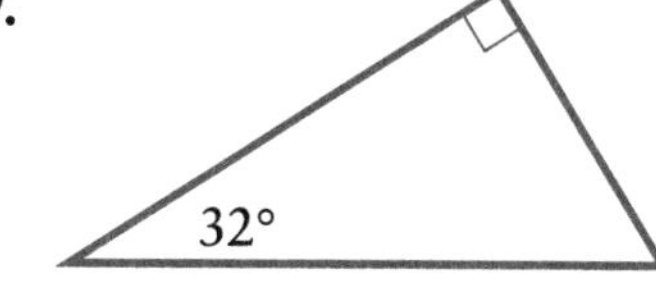

____

10.

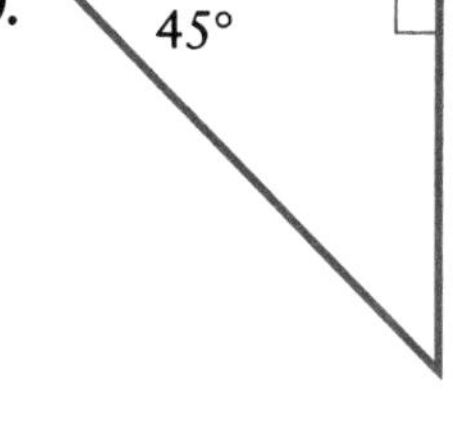

____

11.

____

12. Khoa was doing some tile repair work in the bathroom. The decorative tile all appeared to be right triangles. If one angle measures 40°, what is the measure of the other angle?

To check your answers, turn to page 118.

Pythagorean Theorem

Miranda leans a ladder against the side of her house as in the figure at the right. When viewed from the side, the ladder, house, and ground form a right triangle.

The lengths of the sides in a right triangle have a special relationship, discovered thousands of years ago by a Greek mathematician named Pythagoras. The longest side of a right triangle (the hypotenuse) is c. The other two legs are named a and b. Pythagoras found that:

$$c^2 = a^2 + b^2$$

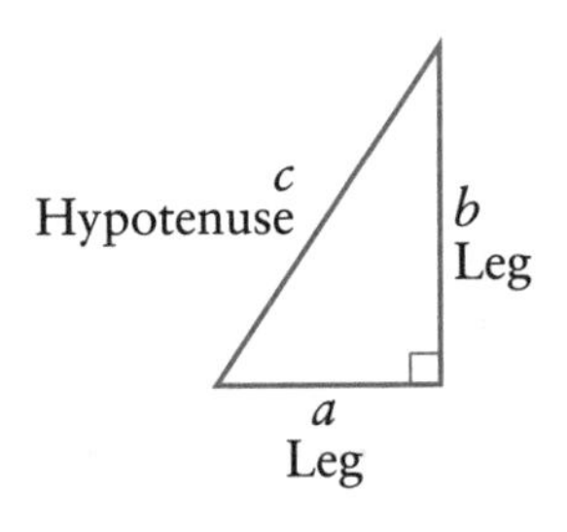

or the "square of the hypotenuse is equal to the sum of the squares of the other two legs." This relationship is called the **Pythagorean theorem.** If you know the lengths of two of the sides in a right triangle, you can use the Pythagorean theorem to find the length of the third side.

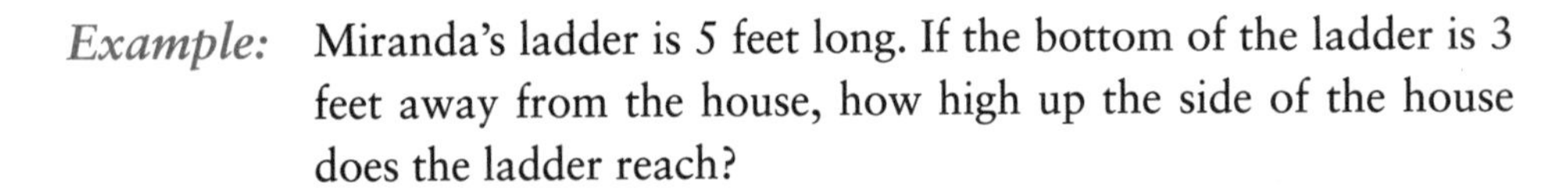

Example: Miranda's ladder is 5 feet long. If the bottom of the ladder is 3 feet away from the house, how high up the side of the house does the ladder reach?

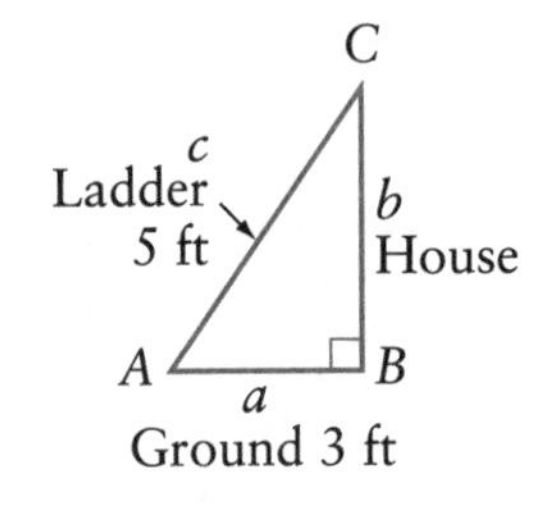

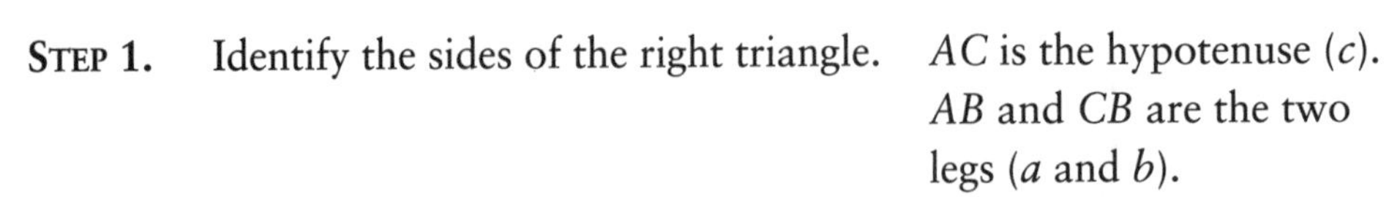

STEP 1. Identify the sides of the right triangle. AC is the hypotenuse (c). AB and CB are the two legs (a and b).

STEP 2. Substitute the lengths of the sides into the formula for the Pythagorean theorem.

$c^2 = a^2 + b^2$
$5^2 = 3^2 + b^2$

STEP 3. Solve the formula.

$5^2 = 3^2 + b^2$
$25 = 9 + b^2$
$25 - 9 = 9 - 9 + b^2$
$16 = b^2$
$\sqrt{16} = \sqrt{b^2}$
$4 = b$

➡ The ladder reaches 4 feet up the side of Miranda's house.

PRACTICE 16

Fill in the blank.

1. The longest side of a right triangle is the ______________.

2. The other two sides of a right triangle are called ______________.

3. The ______________ theorem shows the relationship of the three sides in a right triangle.

4. A right triangle always has one ______________ angle.

5. A right triangle always has two ______________ angles.

Find the missing length.

6. 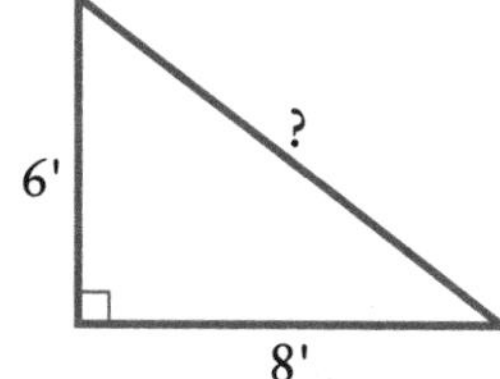

____

7. 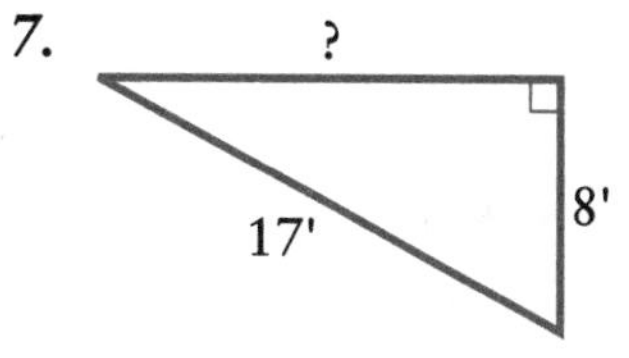

____

8. 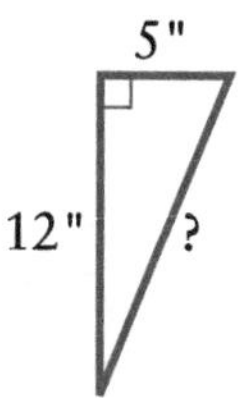

____

9. 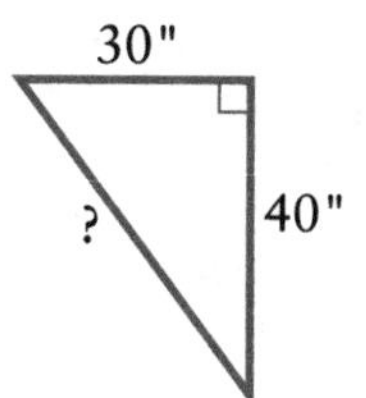

____

Find the missing length. (**Hint:** Drawing a picture may be helpful.)

10. Jo is making a window display, which is shaped as a right triangle. If the legs of the triangle are 6' and 8', what is the length of the longest side? What is the perimeter of the display area?

11. A 25-foot ladder was propped 15 feet from the house. How far up the side of the house did the ladder reach?

12. Kym walked three blocks north and then four blocks west. If she had walked in a straight line, how far would she have walked?

13. Ted is making a shelf to fit a corner in the kitchen. The shelf looks like a right triangle. If the longest side is 13" and the shortest side is 5", how long is the third side?

To check your answers, turn to page 118.

Congruent Triangles

Jawon has two triangular pieces of paper. What measurements does he need to make to prove they are identical?

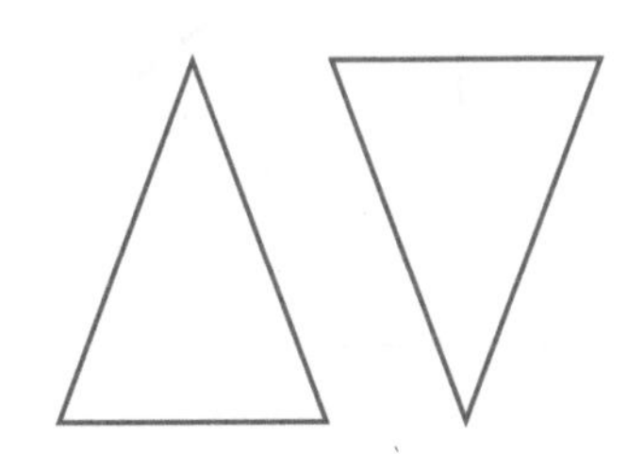

Triangles in which the corresponding angles and sides are equal are called **congruent triangles.** Unlike similar triangles, which have equal corresponding angles and proportional sides, congruent triangles are *identical* figures.

If the sides of both triangles are equal in length, the two triangles are congruent and the corresponding angles must be equal.

Example: Prove these two triangles are congruent.

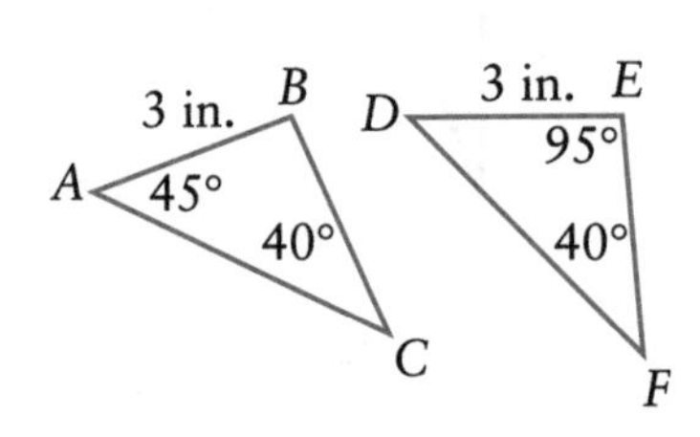

STEP 1. Match the corresponding angles and/or sides.

AB and *DE*	∠*ABC* and ∠ *DEF*
BC and *EF*	∠*BCA* and ∠ *EFD*
AC and *DF*	∠*CAB* and ∠ *FDE*

Step 2. Find the missing angles, if possible.

$\angle BAC = 45°$	$\angle DEF = 95°$
$\angle BCA = 40°$	$\angle EFD = 40°$
$45° + 40° = 85°$	$95° + 40° = 135°$
$180° - 85° = 95°$	$180° - 135° = 45°$
$\angle ABC = 95°$	$\angle EDF = 45°$

Step 3. Find the corresponding angles and/or sides that are equal.

$AB = DE$ $\quad$ $\angle ABC = \angle DEF$
$\angle BCA = \angle EFD$
$\angle CAB = \angle FDE$

➠ Since all three corresponding angles are equal and one pair of corresponding sides is equal, the triangles must be congruent.

PRACTICE 17

Write *yes* if the triangles are congruent; *no* if they are not.

1.

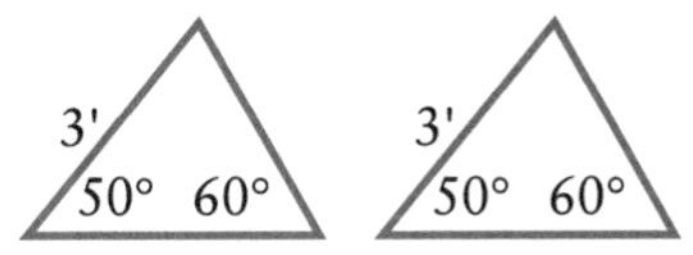

____

2.

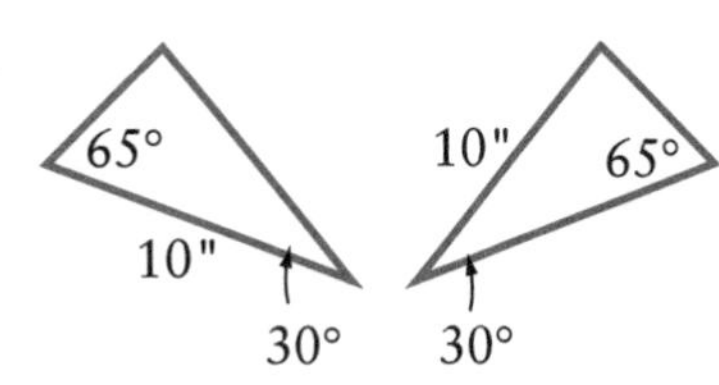

____

3.

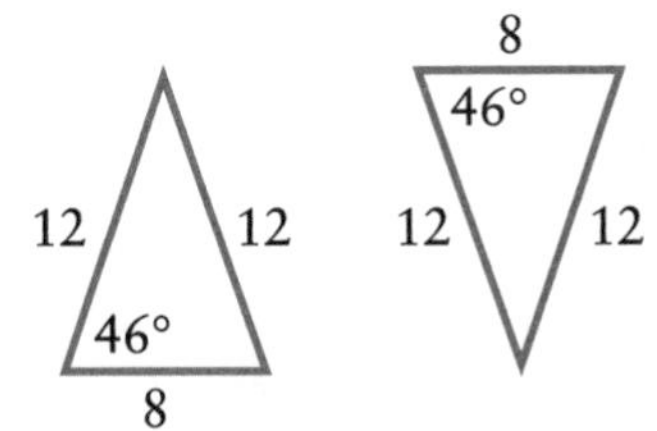

____

4.

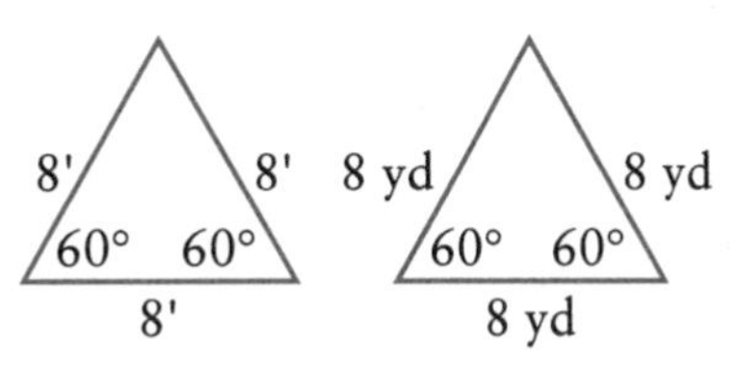

____

5.

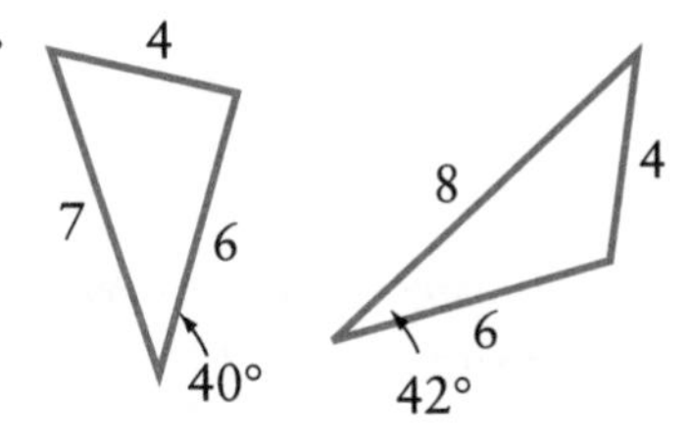

____

6.

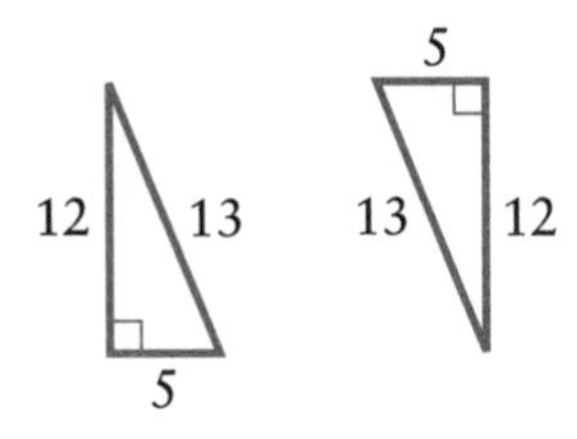

____

Find the measure of the missing side in the congruent triangles.

7.

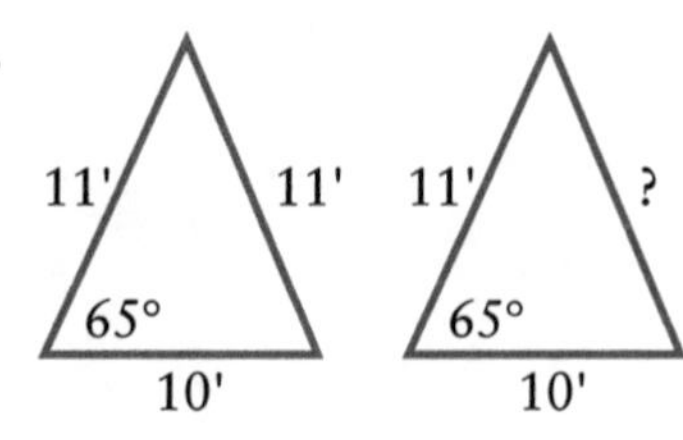

______

8.

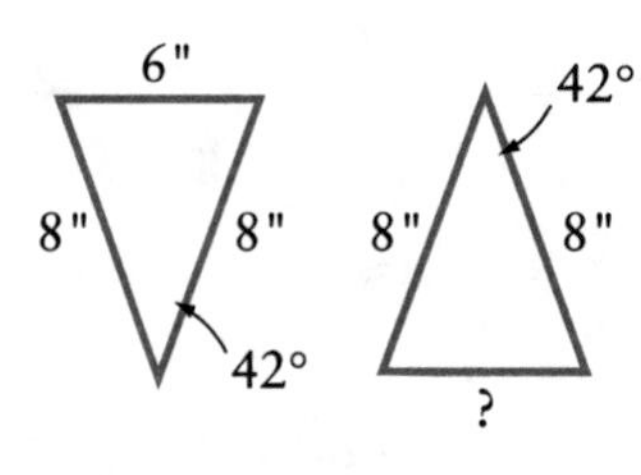

______

Use the following diagram to answer items 9–15.

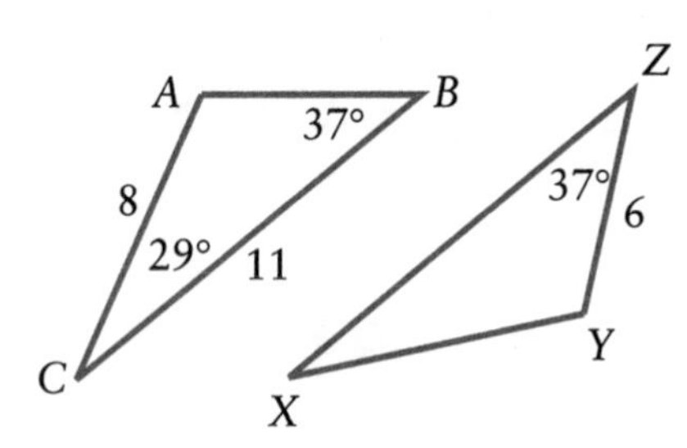

$\triangle ABC$ is congruent to $\triangle YZX$.

9. What are the corresponding angles?

10. What is the length of side XY?

11. What is the length of side AB?

12. What is the length of side XZ?

13. What is the measure of $\angle X$?

14. What is the measure of $\angle A$?

15. What is the measure of $\angle Y$?

To check your answers, turn to page 119.

TRIANGLES REVIEW

The problems on pages 51 to 53 will help you find out if you need to review the triangles section of this book. When you finish, look at the chart on page 53 to see which pages you should review. Fill in the blank.

1. The sum of the angles in a triangle is ______________°.

2. A triangle has ______________ sides.

3. The formula for the area of a triangle is ______________.

4. The formula for the perimeter of a triangle is ______________.

5. The ______________ theorem shows the relationship of the sides in a right triangle.

Match the two columns.

_______ *6.* similar triangles *a.* all sides different

_______ *7.* right triangle *b.* two sides equal

_______ *8.* equilateral triangle *c.* same size, same shape

_______ *9.* congruent triangles *d.* same shape, different size

_______ *10.* isosceles triangle *e.* has one right angle

_______ *11.* scalene triangle *f.* all sides equal

What kind of triangle is each?

12. 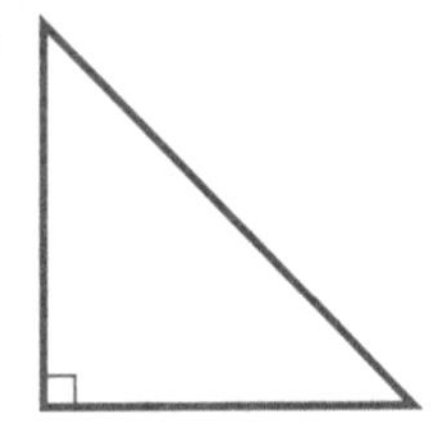

13. 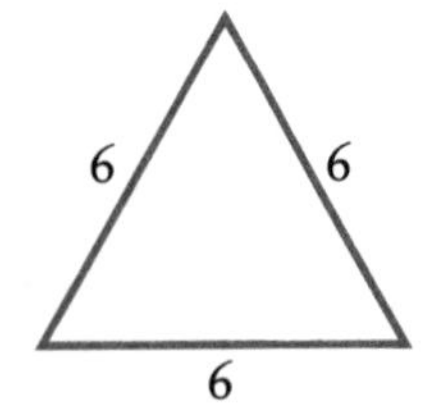

14. 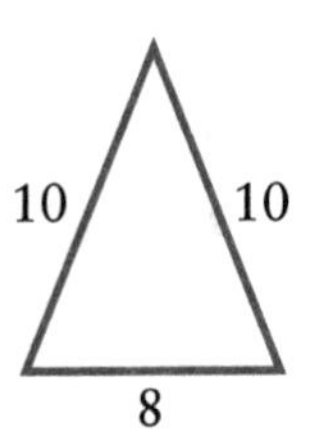

__________ __________ __________

15. 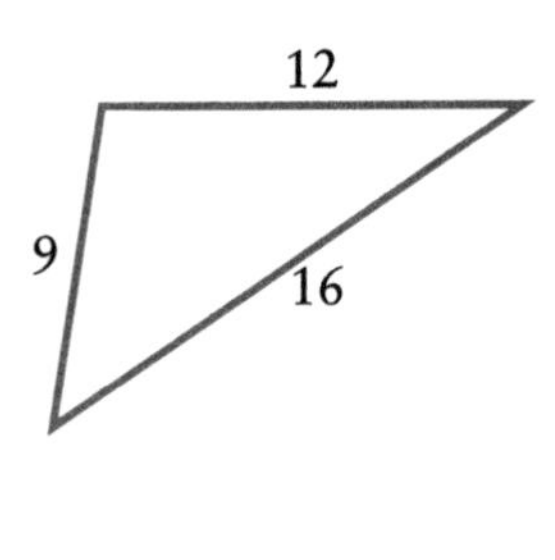

16. 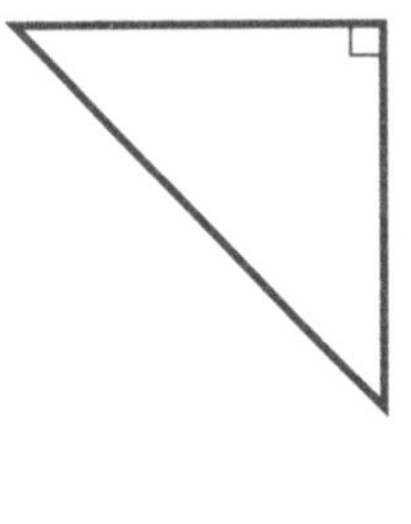

__________ __________

Find the measure of the missing angle.

17. 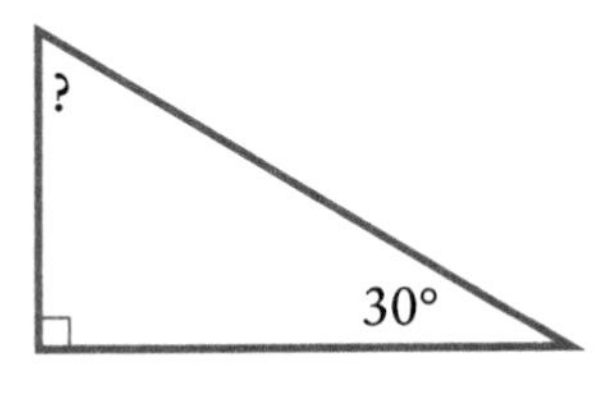

18. 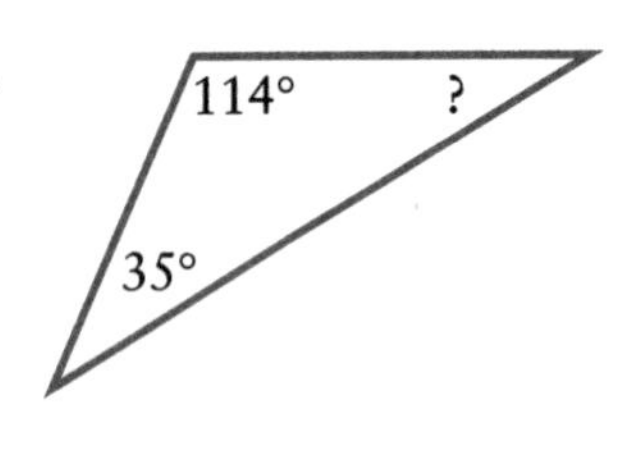

________ ________

Find the area and perimeter of each triangle.

19. 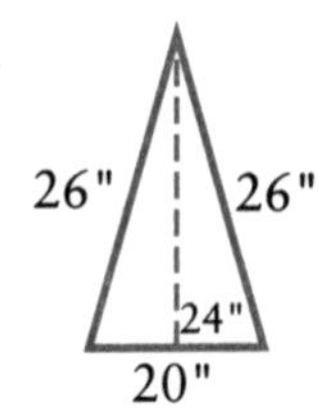

$A =$ ________

$P =$ ________

20. 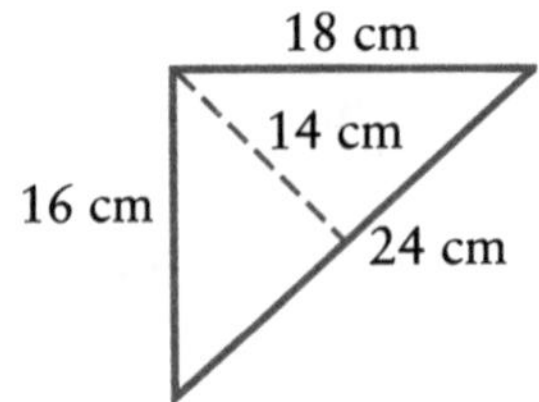

$A =$ ________

$P =$ ________

21. 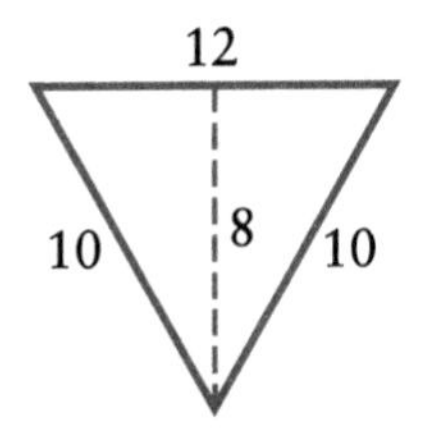

$A =$ ________

$P =$ ________

22. 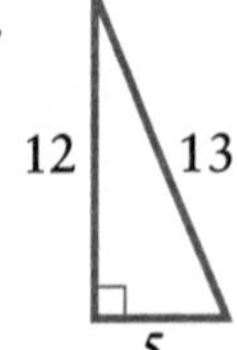

$A =$ ________

$P =$ ________

Find the measure of the missing side.

23.

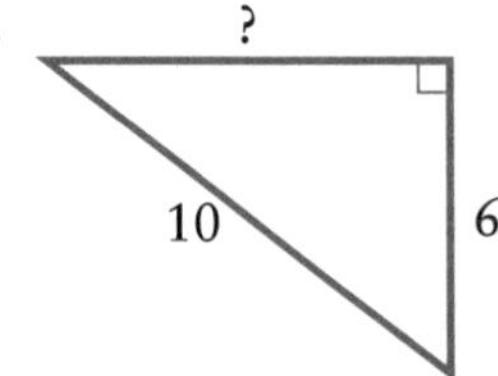

____

24.

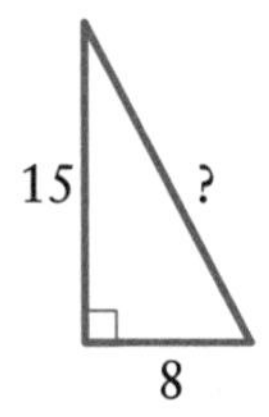

____

25.

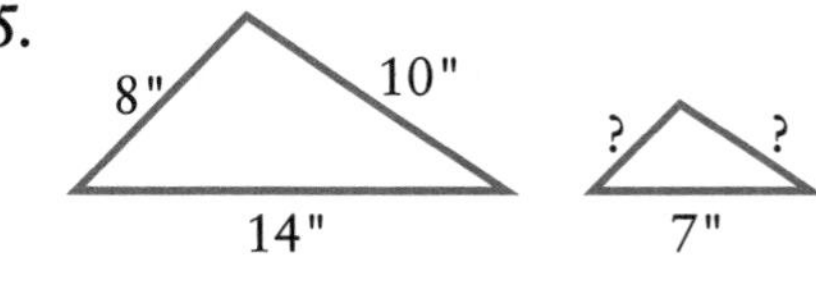

____

26. Jason drove 5 miles and then turned right. He drove 12 more miles. If he had driven in a straight line, how far would he have driven?

27. If the perimeter of an equilateral triangle is 60 inches, what is the length of one side?

PROGRESS CHECK

Check your answers on page 119. Then return to the review pages for the problems you missed. Correct your answers before going on to the next unit.

If you missed problems	*Review pages*
2, 7–8, 10–16	29 to 32
1, 17–18	32 to 33
3–4, 19–22	34 to 36
6, 25	36 to 38
27	38 to 40
5, 23–24, 26	46 to 48
9	48 to 50

UNIT 3

Quadrilaterals

Kinds of Quadrilaterals

A **quadrilateral** is a four-sided shape. There are four kinds of quadrilaterals discussed in this unit. A **square** is a quadrilateral with all four sides equal in length and four 90° angles. A **rectangle** also has four 90° angles, but only opposite sides must be equal in length. Both pairs of opposite sides of a **parallelogram** are parallel and equal in length. In a **trapezoid,** only one pair of opposite sides is parallel. Nonparallel sides may or may not be equal. The sum of the angles of any quadrilateral equals 360°.

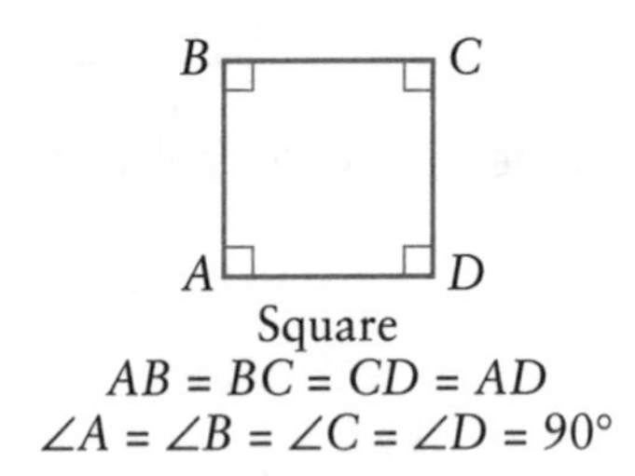

Square
$AB = BC = CD = AD$
$\angle A = \angle B = \angle C = \angle D = 90°$

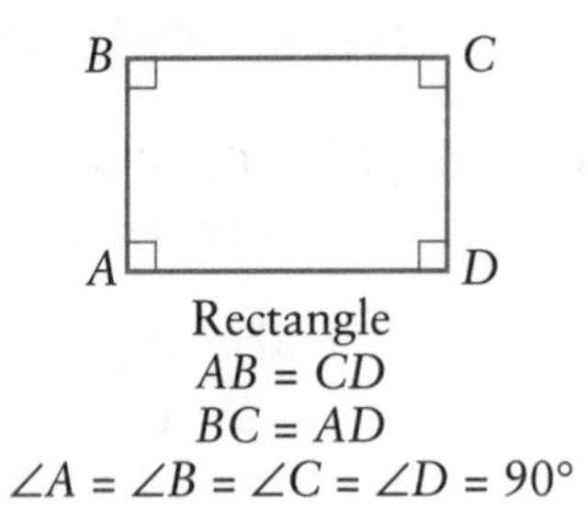

Rectangle
$AB = CD$
$BC = AD$
$\angle A = \angle B = \angle C = \angle D = 90°$

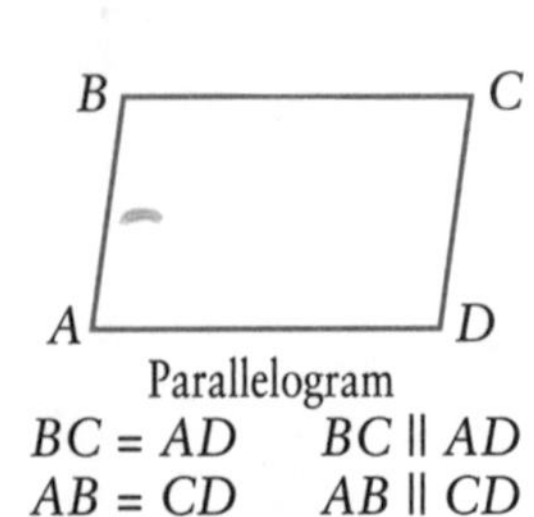

Parallelogram
$BC = AD$ $BC \parallel AD$
$AB = CD$ $AB \parallel CD$

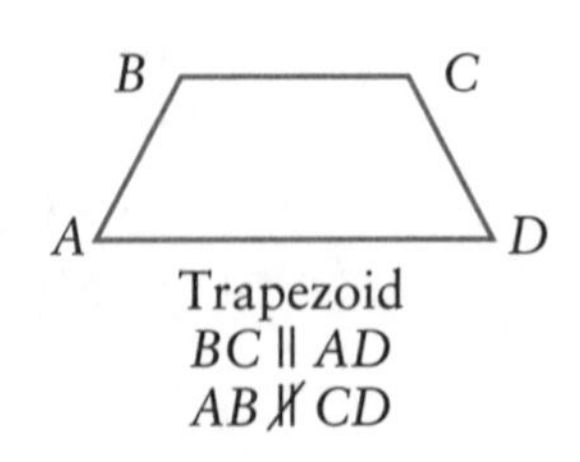

Trapezoid
$BC \parallel AD$
$AB \nparallel CD$

Martina is writing to the company who makes her son's building blocks. Her son is missing one of the pieces. In describing the missing piece, she notes that it is a four-sided block.

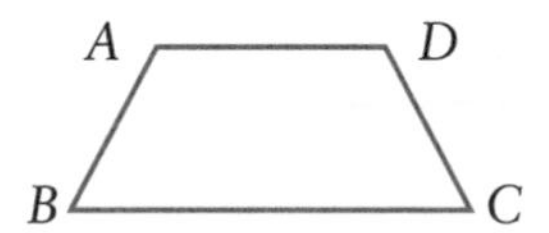

Example: Determine the kind of quadrilateral block Martina's son is missing.

STEP 1. Measure the lengths of the sides of the quadrilateral. $AB = DC; AD \neq BC$

STEP 2. Determine which, if any, of the opposite sides are parallel. $BC \parallel AD; AB \nparallel DC$

STEP 3. Determine which, if any, of the angles is 90°. none of the angles is 90°

STEP 4. Find the kind of quadrilateral, using the table below:

Description	*Type of Quadrilateral*
4 equal sides, four 90° angles	square
2 opposite pairs of equal sides, four 90° angles	rectangle
2 opposite pairs of equal parallel sides	parallelogram
1 opposite pair of parallel sides	trapezoid

➡ The missing block is a trapezoid.

PRACTICE 18

Label each quadrilateral.

1.

____

2.

____

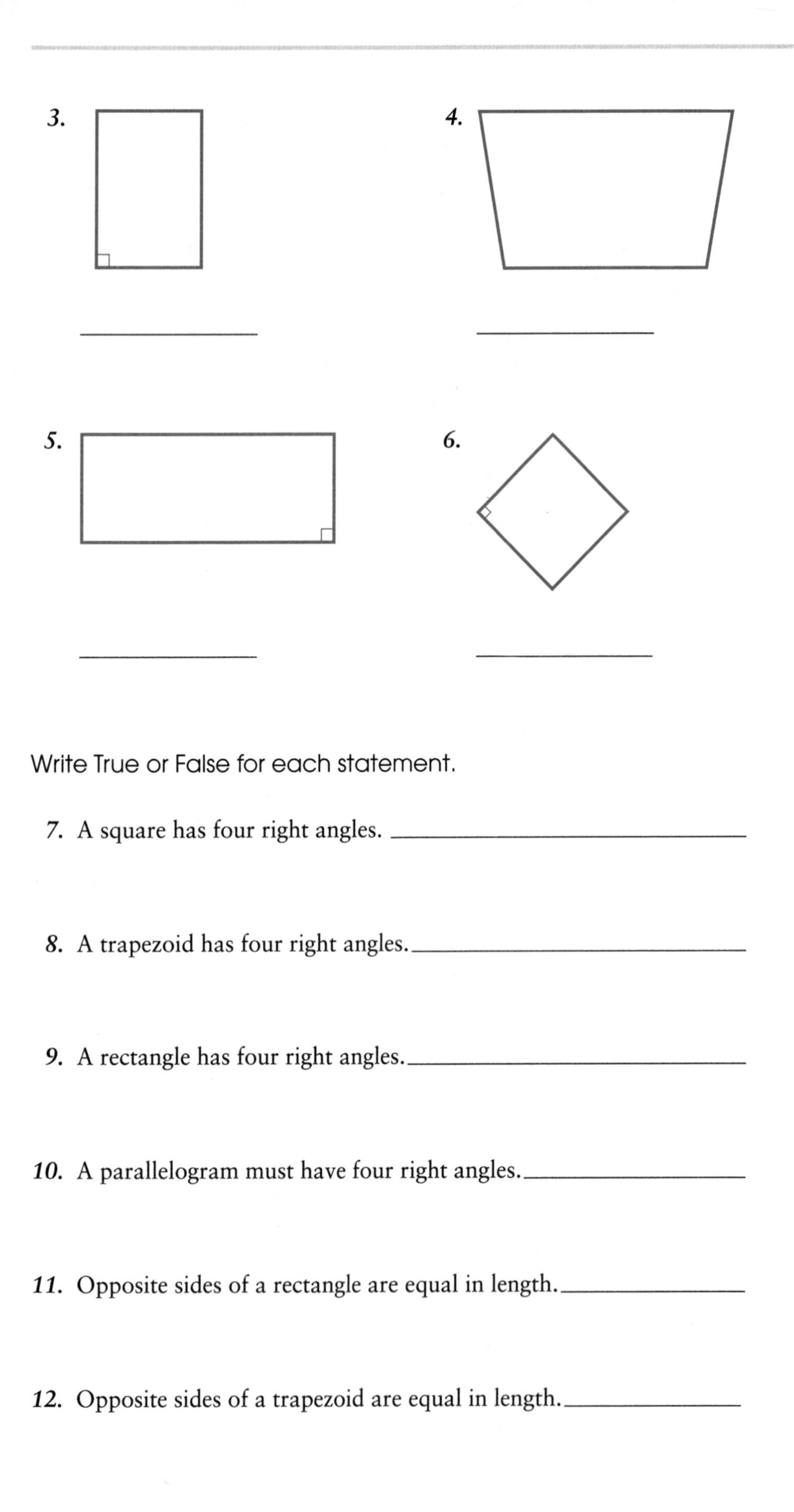

Write True or False for each statement.

7. A square has four right angles. ______________________

8. A trapezoid has four right angles. ______________________

9. A rectangle has four right angles. ______________________

10. A parallelogram must have four right angles. ______________

11. Opposite sides of a rectangle are equal in length. __________

12. Opposite sides of a trapezoid are equal in length. __________

Find the measure of the indicated angles.

13. 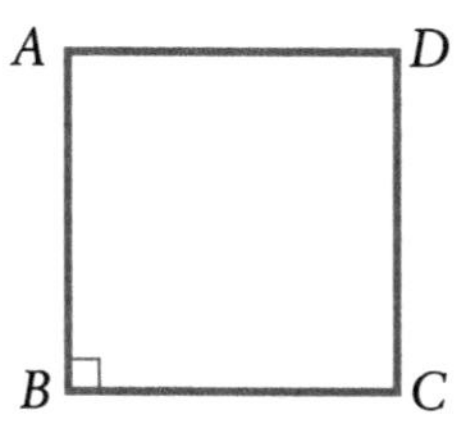

$\angle A =$ ________

Square

14. 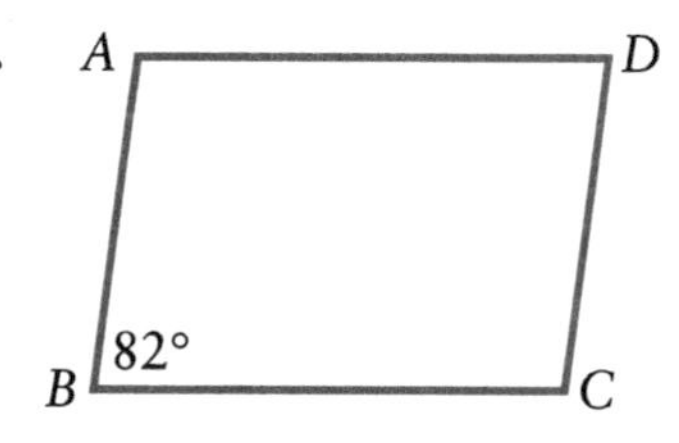

$\angle A =$ ________

$\angle C =$ ________

Parallelogram

15. 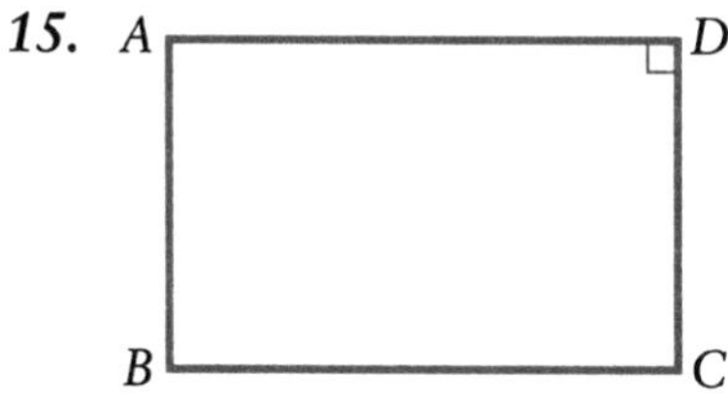

$\angle B =$ ________

Rectangle

16. 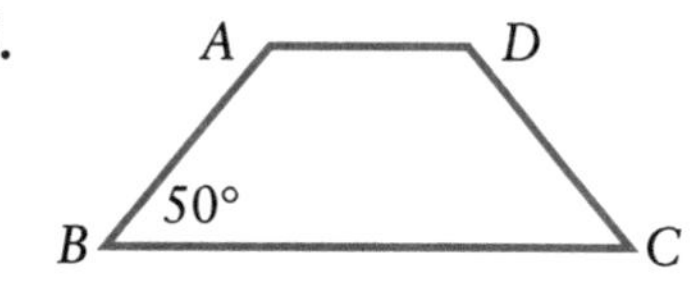

$\angle A =$ ________

Trapezoid

To check your answers, turn to page 120.

Perimeter and Area of Squares

Marco wants to place a square piece of carpeting on his living room floor. He also needs to place double-stick tape under the outside edge of the carpeting to keep it in place.

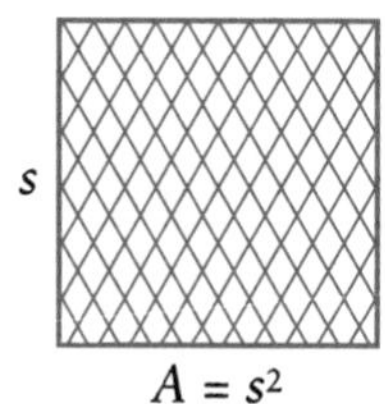

$A = s^2$

The **area** is the surface amount in a figure. Area is measured in square units. A square is a quadrilateral with four equal sides and four right angles. The area (A) of a square can be found by multiplying the length of two sides (s). The formula is:

$$A = s \times s \quad \text{or} \quad A = s^2$$

The *perimeter* is the distance around a figure. The perimeter is measured in linear units. Since a square has four equal sides, the perimeter (P) can be found by adding the length of a side (s) four times or multiplying the length by four.

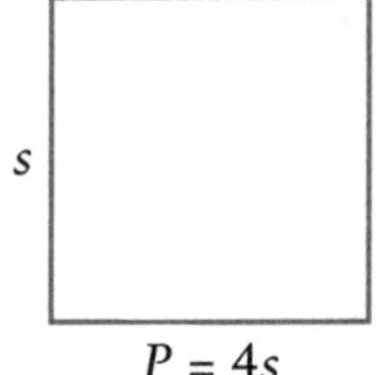

$$P = s + s + s + s \quad \text{or} \quad P = 4s$$

Example: Determine the area of the square carpet Marco needs to buy, if one side is 5 ft.

STEP 1. Find the length of one side of the square. $s = 5$ feet

STEP 2. Substitute the length in the formula. $A = s^2 = 5^2$
$A = 25$ square feet

➡ Marco needs 25 square feet of carpeting.

Example: Find how much tape Marco will need to attach the outside border of the carpet to the floor.

STEP 1. Find the length of one side of the square. $s = 5$ feet

STEP 2. Substitute the length in the formula. $P = 4s = 4 \times 5 = 20$ feet

➡ Marco will need 20 feet of tape to attach the carpet to the floor.

PRACTICE 19

Find the area and perimeter of each square.

1. 4'
$A =$ ________
$P =$ ________

2. 2.5 m
$A =$ ________
$P =$ ________

3. 1 yd
$A =$ ________
$P =$ ________

4. 6.5'
$A =$ ________
$P =$ ________

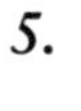

5. 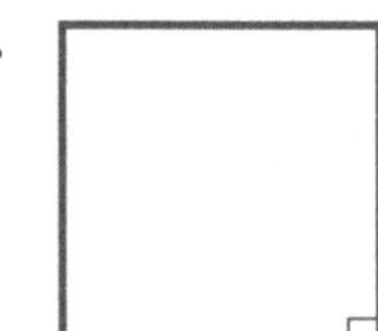

80 blocks

$A =$ ________

$P =$ ________

6. 16"

$A =$ ________

$P =$ ________

7. 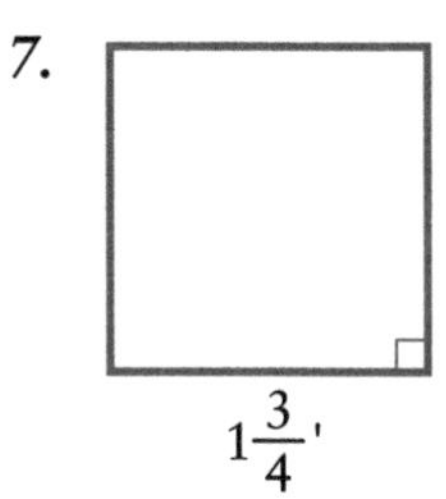

$1\frac{3}{4}'$

$A =$ ________

$P =$ ________

8. 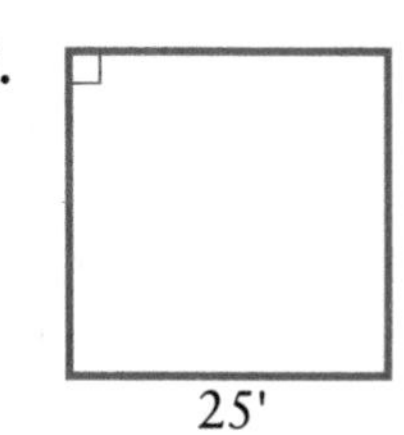

25'

$A =$ ________

$P =$ ________

9. Margo wanted to put weather stripping around the kitchen windows. There are four square windows of 30" on a side. How much weather stripping does she need?

10. Harold is making square picture frames. If the side of a frame is 8", how much wood does he need for one frame? How much wood does he need for five frames?

11. Rafael wants to put contact paper on an 18" square table top. What is the area of the table top?

12. Horatio has a square corn field. If one side is 250 yards long, what is the area of the field?

13. Maria wants to cover a square pillow with new fabric and trim. If the pillow is 14" on a side, what is the perimeter of the pillow? What is the area of the pillow?

To check your answers, turn to page 120.

Perimeter and Area of Rectangles

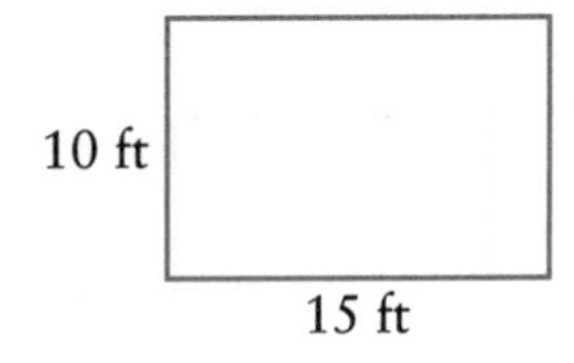

Julia has a wall that she would like to decorate. The wall is rectangular. Julia wants to paint the wall, then put a border around the edges.

The area (A) of a rectangle can be found by multiplying the length (l) times the width (w) in the formula: $A = lw$. The perimeter (P) of a rectangle is two times the length (l) plus two times the width (w) in the formula: $P = 2l + 2w$.

Example: Find the area of Julia's 15'-by-10' rectangular wall.

STEP 1. Write the length and width of the rectangle.

$l = 15$ feet
$w = 10$ feet

STEP 2. Substitute the numbers in the formula and solve.

$A = lw$
$A = 15 \times 10$
$A = 150$ square feet

➡ Julia needs enough paint to cover 150 square feet.

Example: Find the length of a border around the edges of Julia's wall.

STEP 1. Write the length and width of the rectangle.

$l = 15$ feet
$w = 10$ feet

STEP 2. Substitute the numbers in the formula and solve.

$P = 2l + 2w$
$P = 2 \times 15 + 2 \times 10$
$P = 30 + 20$
$P = 50$ feet

➡ Julia needs 50 feet of border to surround the wall.

PRACTICE 20

A. Find the area and perimeter of each rectangle.

1.

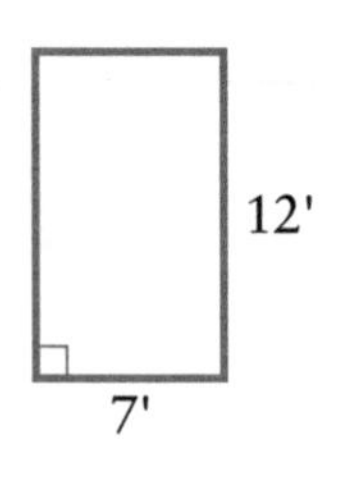

$A =$ ________

$P =$ ________

2.

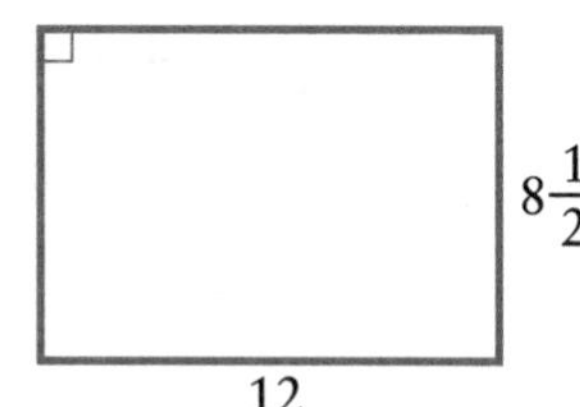

$A =$ ________

$P =$ ________

3.

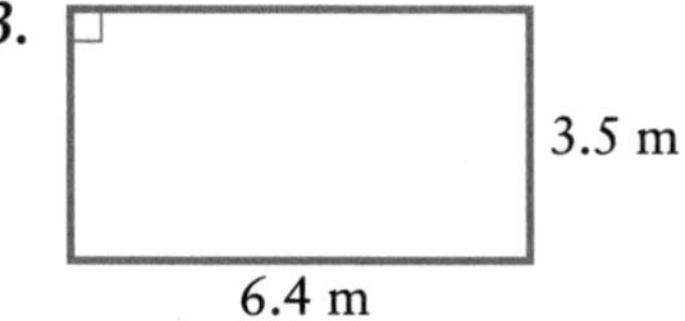

$A =$ ________

$P =$ ________

4.

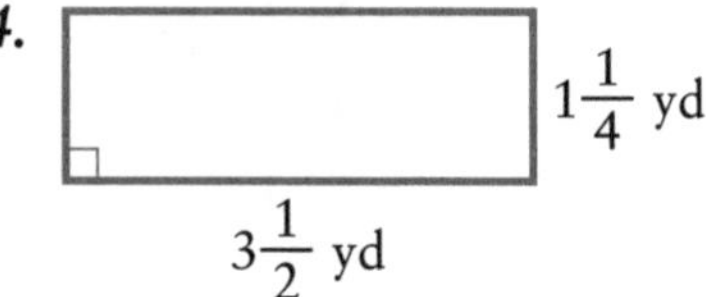

$A =$ ________

$P =$ ________

5.

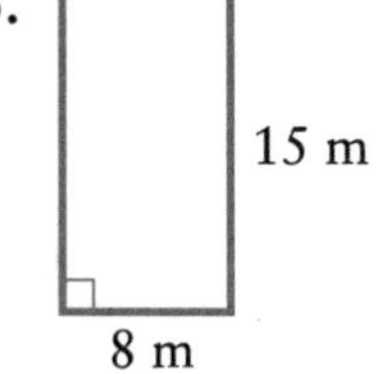

$A =$ ________

$P =$ ________

6.

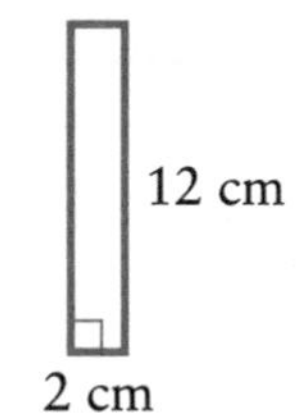

$A =$ ________

$P =$ ________

Sometimes measurements are not given in the same units.

Example: Find the perimeter of this rectangle.

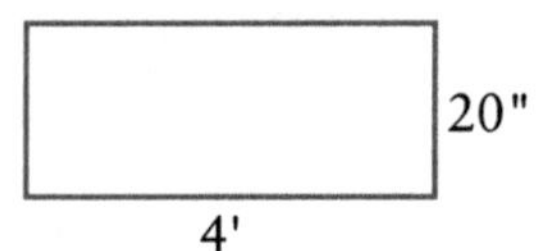

STEP 1.	Convert one measurement.	$4' = 4 \times 12" = 48"$
STEP 2.	Substitute the numbers.	$P = 2 \times 48 + 2 \times 20$
STEP 3.	Complete the problem.	$P = 96 + 40 = 136"$

➠ The perimeter is 136 inches.

B. Find the area and perimeter of each. Convert one measurement, if necessary.

7.

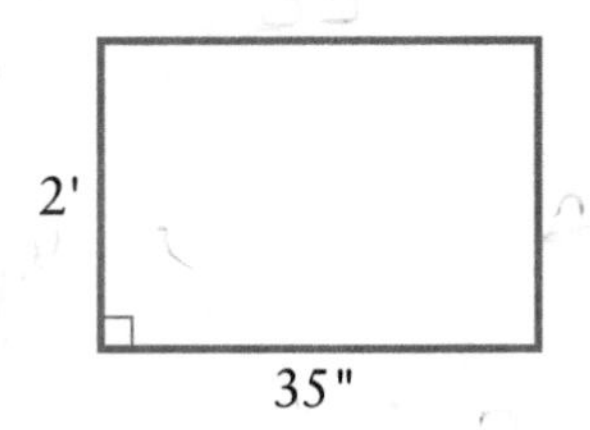

$A =$ ________

$P =$ ________

8.

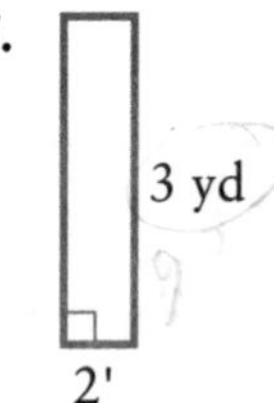

$A =$ ________

$P =$ ________

9.

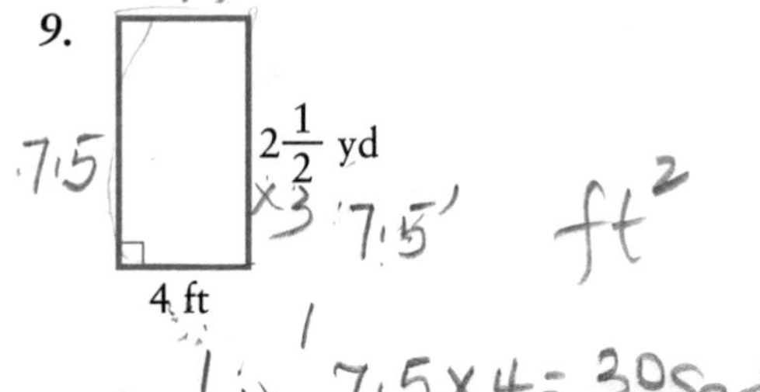

$A =$ ________

$P =$ ________

10.

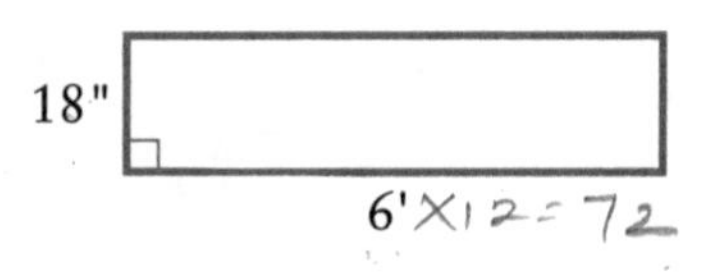

$A =$ ________

$P =$ ________

11.

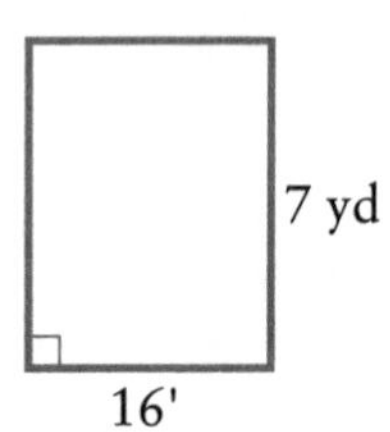

$A =$ ________

$P =$ ________

12.

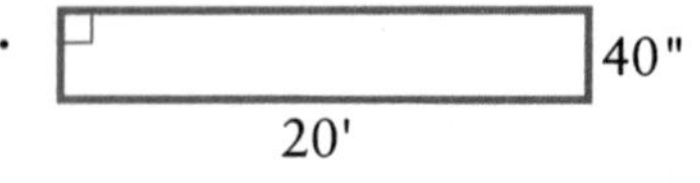

$A =$ ________

$P =$ ________

13. Jose wants to buy new carpet for his rectangular living room. The dimensions are 14 feet by 4 yards. What is the area of the room?

14. Jose also wants to put new wood trim around the ceiling. What is the perimeter of the room?

To check your answers, turn to page 121.

Perimeter and Area of Parallelograms

Don's front yard is in the shape of a parallelogram. Don is buying grass seed to plant the lawn. After planting, he plans to enclose the lawn with wire fencing. To know how much seed to buy, Don needs to find the area of his lawn. To know how much fencing to buy, he needs to know the perimeter of the lawn.

A parallelogram is a four-sided figure, with parallel opposite sides. To find the area (A) of a parallelogram, multiply the base times the height. The **base** (b) is the length of one side. The **height** (h) is the distance from the base to the top of the parallelogram.

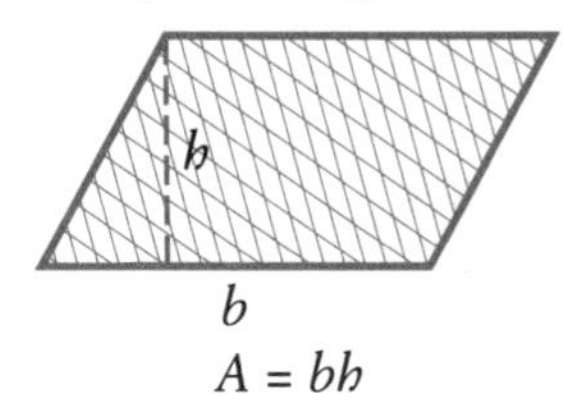

$A = bh$

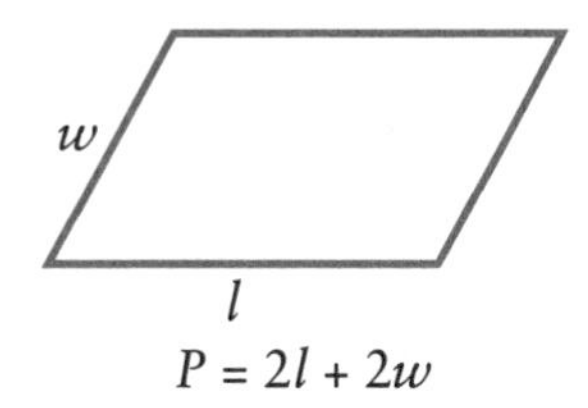

$P = 2l + 2w$

Example: Find the area of Don's yard, based on the diagram.

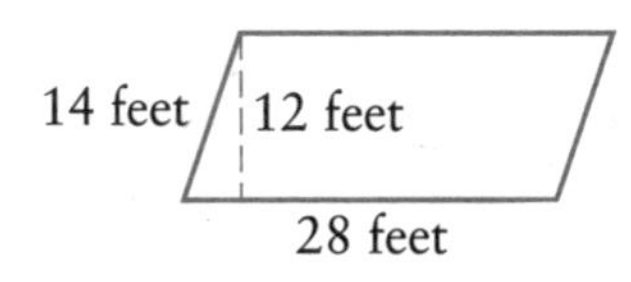

STEP 1. Write the base and height of the parallelogram.

$b = 28$ feet
$h = 12$ feet

STEP 2. Substitute the numbers in the formula.

$A = bh$
$A = 28 \times 12$
$A = 336$ square feet

You do not need the width (14 ft) to find area.

➠ Don needs enough grass seed to cover 336 square feet of lawn.

To determine the perimeter (P), use the same formula as for a rectangle, two times the length (l) plus two times the width (w):

$$P = 2l + 2w$$

Example: Find the perimeter of Don's yard.

The length of a parallelogram is equal to the base.

STEP 1. Write the length and width of the parallelogram.

$l = 28$ feet
$w = 14$ feet

STEP 2. Substitute the numbers in the formula and solve for P.

$P = 2l + 2w$
$P = 2(28) + 2(14)$
$P = 56 + 28$
$P = 84$ feet

➡ Don needs 84 feet of wire fencing to enclose the lawn.

PRACTICE 21

A. Write *yes* if the figure is a parallelogram; *no* if it is not.

1.

____

2.

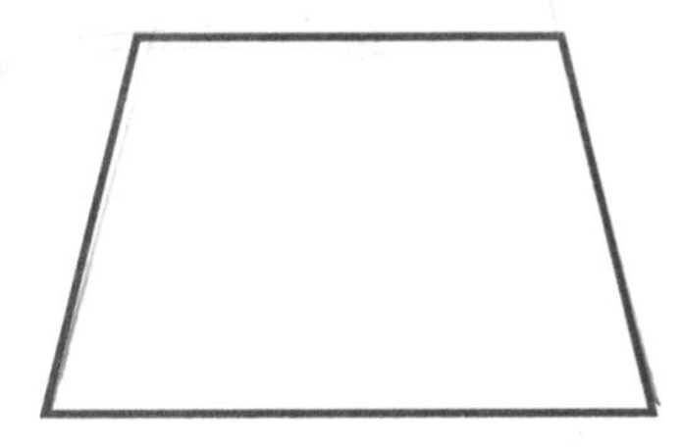

____

3.

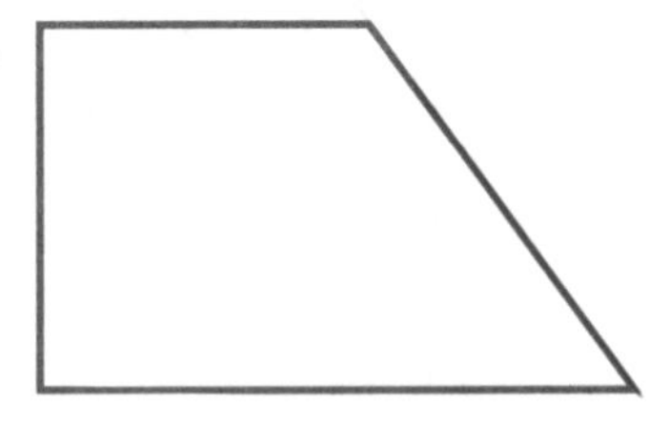

____

4.

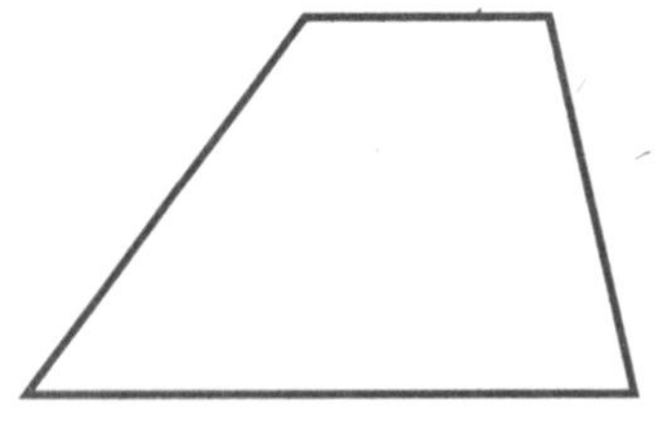

____

Find the perimeter and area of each parallelogram.

5.

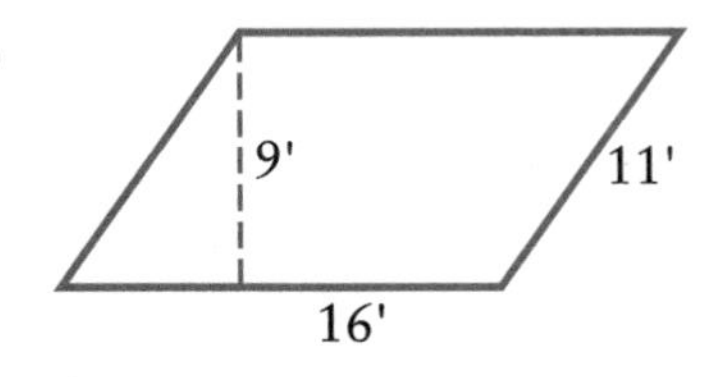

$A =$ ________

$P =$ ________

6.

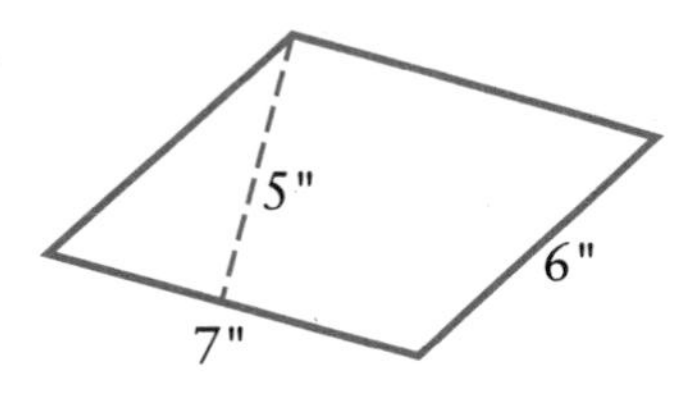

$A =$ ________

$P =$ ________

7.

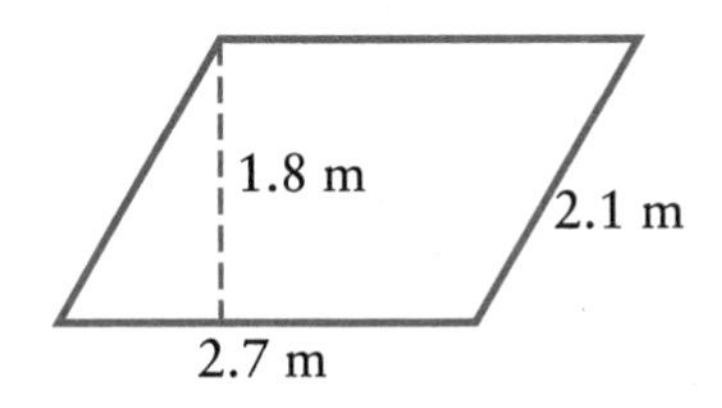

$A =$ ________

$P =$ ________

8.

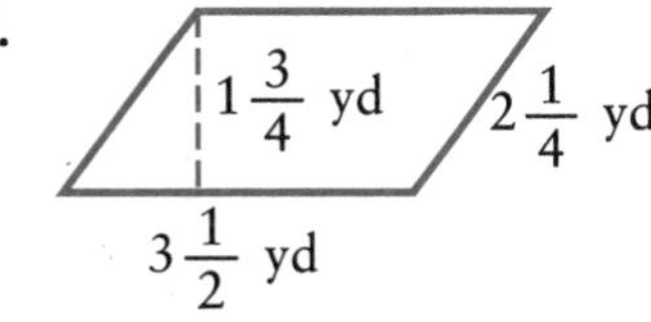

$A =$ ________

$P =$ ________

Sometimes measurements are not given in the same units.

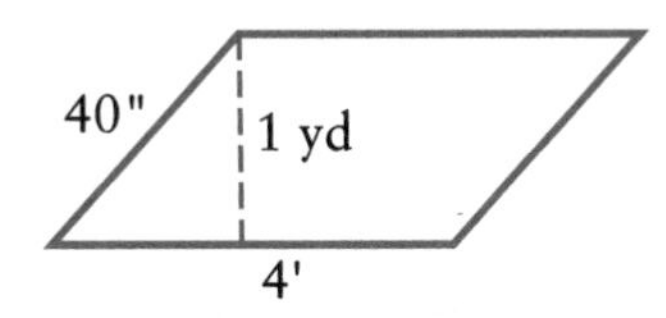

Example: Find the area of the parallelogram.

STEP 1.	Convert one measurement.	1 yd = 3 ft
STEP 2.	Substitute the numbers.	$A = 3 \cdot 4$
STEP 3.	Complete the problem.	$A = 12$ sq ft

➡ The area of the parallelogram is 12 sq ft.

B. Find the area and perimeter of each figure. Convert one measurement, if necessary.

9.

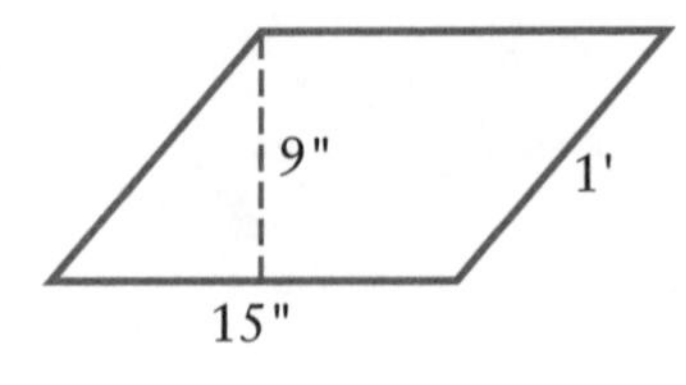

$A =$ ________

$P =$ ________

10.

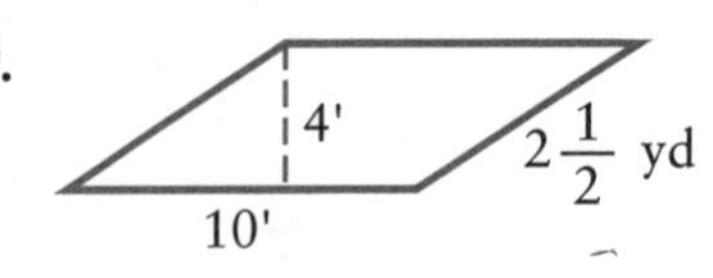

$A =$ ________

$P =$ ________

11.

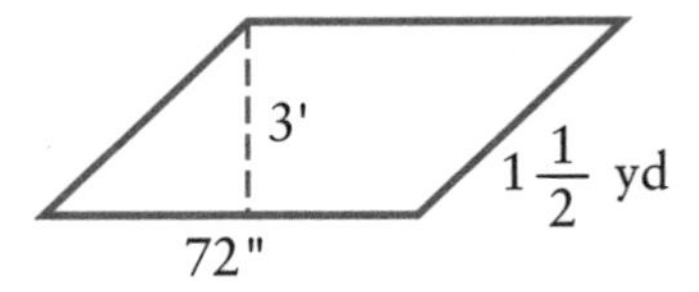

$A =$ ________

$P =$ ________

12.

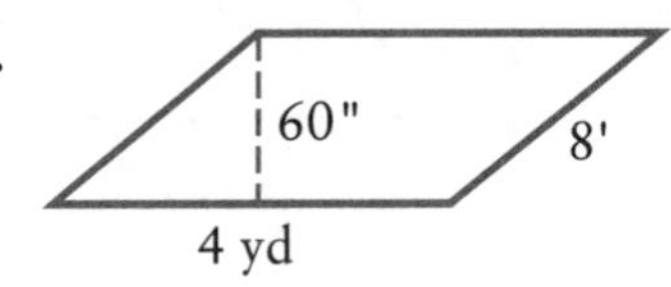

$A =$ ________

$P =$ ________

Use the following information to answer items 13–14.

Abdul was helping a youth group paint a mural on an abandoned building. A special paint in quart size will cover 75 sq feet. The parallelogram to be covered is shown here.

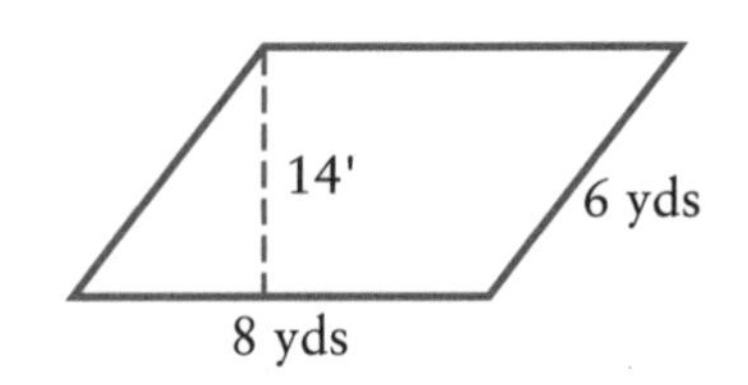

13. How many gallons of paint will Abdul need to cover the parallelogram?

14. What is the perimeter of the parallelogram?

To check your answers, turn to page 122.

QUADRILATERALS REVIEW

The problems on pages 67 to 69 will help you find out if you need to review the quadrilaterals section of this book. When you finish, look at the chart on page 69 to see which pages you should review.

What type of quadrilateral is each?

1.

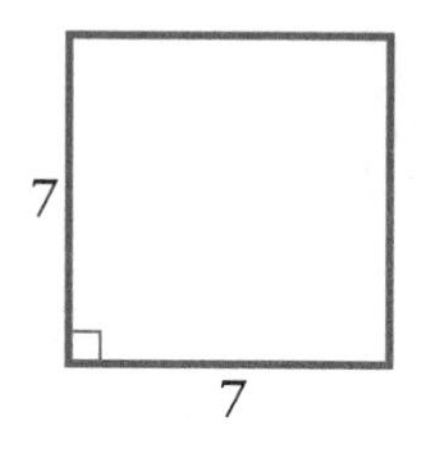

____

2.

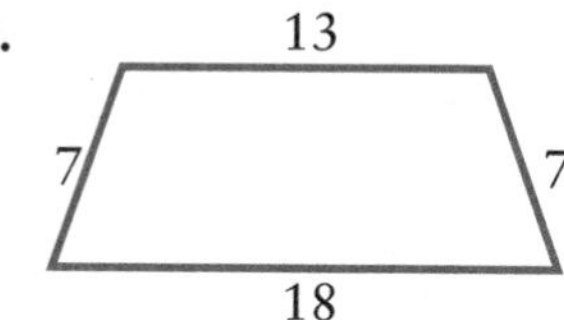

____

3.

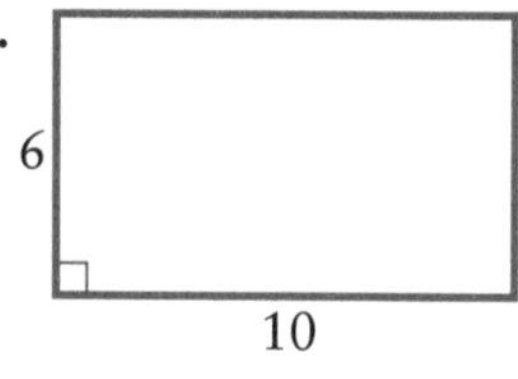

____

4.

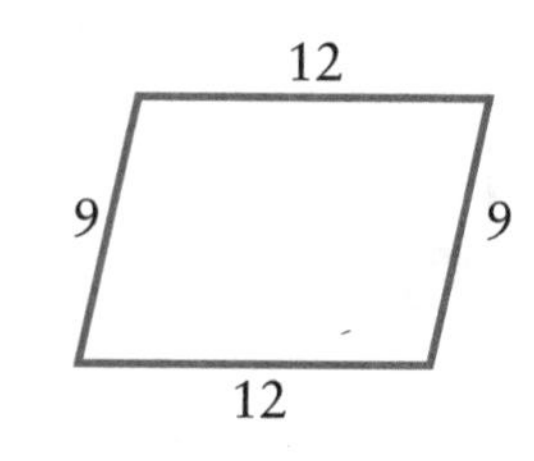

____

5.

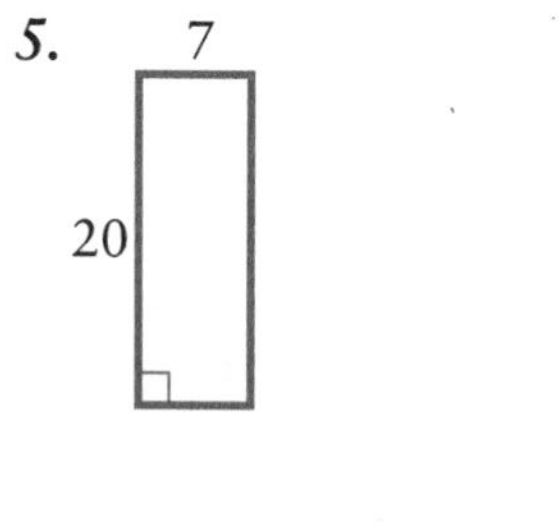

____

6.

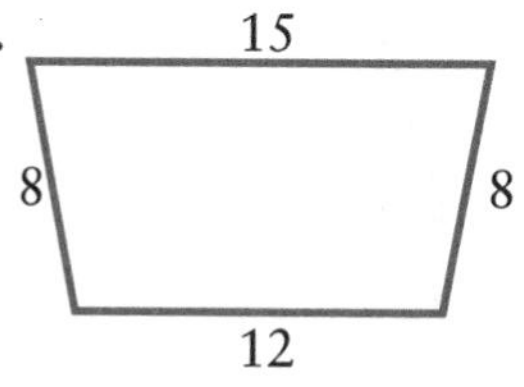

____

Find the area and perimeter of each.

7.

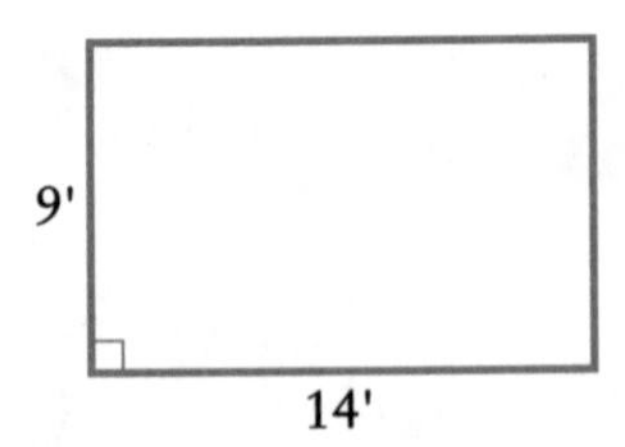

$A =$ ________

$P =$ ________

8.

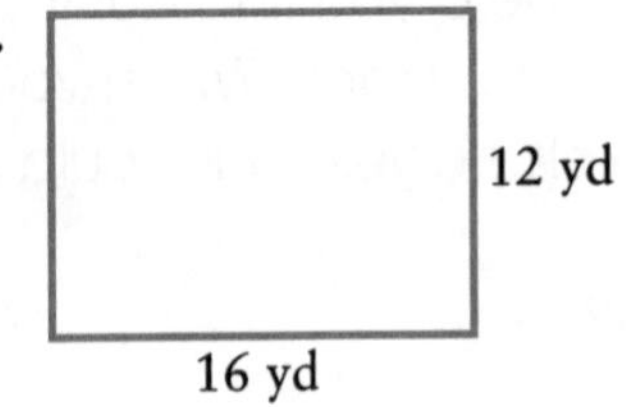

$A =$ ________

$P =$ ________

9.

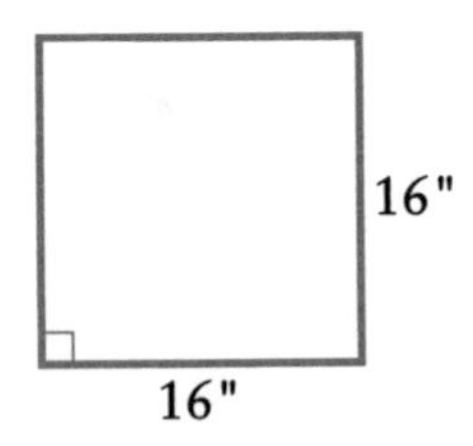

$A =$ ________

$P =$ ________

10.

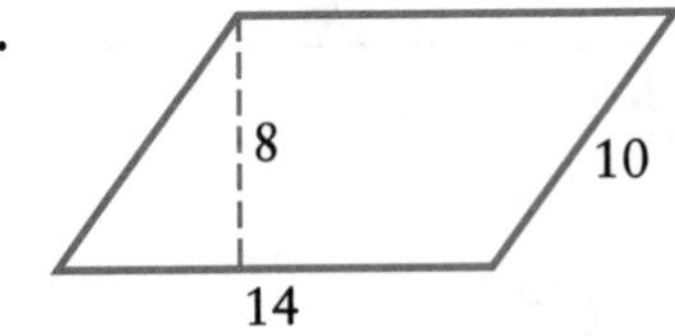

$A =$ ________

$P =$ ________

11.

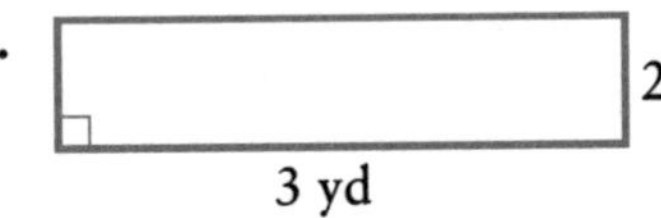

$A =$ ________

$P =$ ________

12.

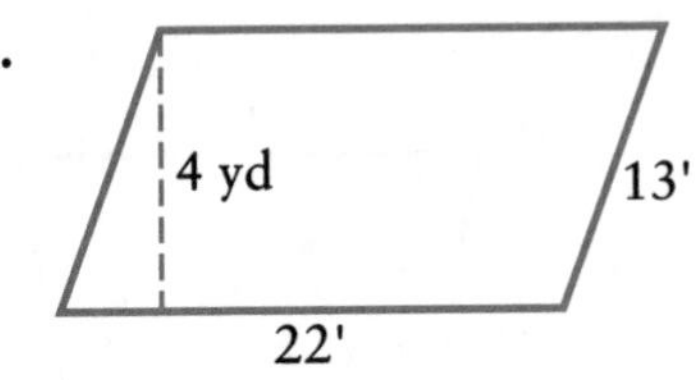

$A =$ ________

$P =$ ________

13.

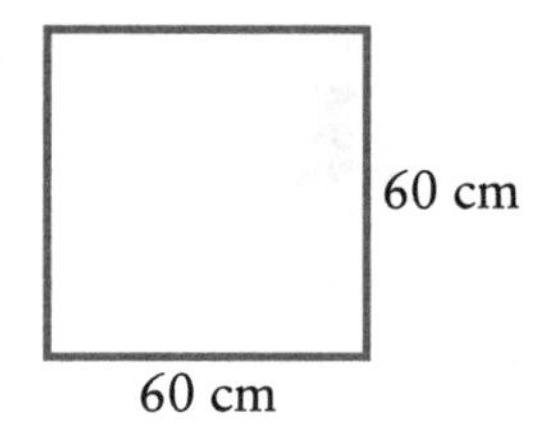

$A =$ ________

$P =$ ________

14.

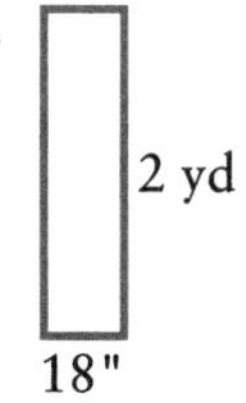

$A =$ ________

$P =$ ________

15. Charo wanted to put weather stripping around three window frames. Each window measured 24" by $3\frac{1}{2}'$. How much weather stripping does she need?

16. The new tile pieces that Samantha wanted on her kitchen counter came in 1" squares. Her counter is 6' long and 20" wide. The store has only 1,500 tiles. Is this enough to complete her counter?

PROGRESS CHECK

Check your answers on page 123. Then return to the review pages for the problems you missed. Correct your answers before going on to the next unit.

If you missed problems	*Review pages*
1 to 6	54 to 57
7 to 8, 11, 14 to 16	60 to 63
9 or 13	58 to 59
10 or 12	63 to 66

UNIT 4

Circles

Finding Circumference

Ngyuen exercises by jogging around a circular track. How far does Ngyuen travel each time she runs around the track?

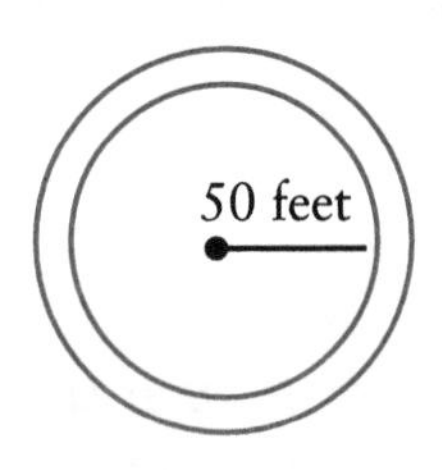

The distance around the edge of a circle is called the **circumference.** To find the circumference (C) of a circle, you must know its radius or diameter. The **radius** of a circle is the distance from the center of the circle to the edge. The **diameter** is the distance of a straight line that divides the circle into two equal parts. The diameter is 2 times the radius.

The circumference of a circle is slightly more than 3 times the diameter. To be more precise, the Greek letter π (*pi*) is used. Rounded to two decimal places, π is about 3.14.

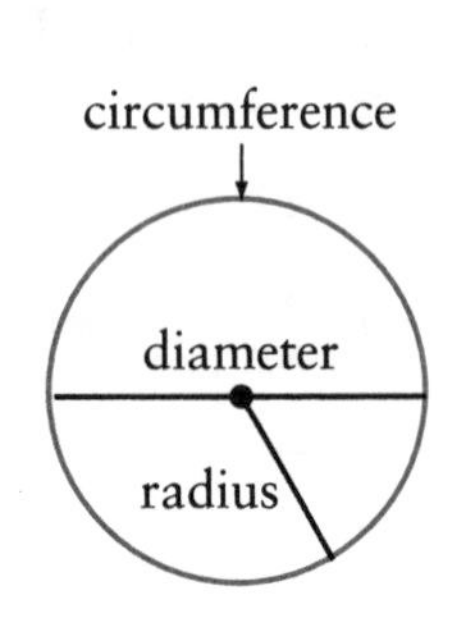

Since the diameter (d) is two times the radius (r), the formula for the circumference (C) of a circle is:

$$C = 2\pi r \quad \text{or} \quad C = \pi d.$$

Example: Compute the distance Ngyuen runs around the track if the distance from the center of the track to the track is 50 feet.

STEP 1. Write the radius of the circle. $r = 50$ feet

STEP 2. Substitute the number in the formula. (Use 3.14 for the value of π.)

$C = 2\pi r$
$C = 2 \times 3.14 \times 50$
$C = 6.28 \times 50$
$C = 314$ feet

➡ Ngyuen runs 314 feet around the track.

PRACTICE 22

Fill in the blanks.

1. The distance around a circle is called the ______________________.

2. The distance from the center of the circle to the edge is the

   ______________.

3. The ______________ divides a circle into two equal parts.

4. The symbol π is equal to ______________.

5. The formula to find the circumference of a circle is ______________.

Find the diameter of each circle.

6. 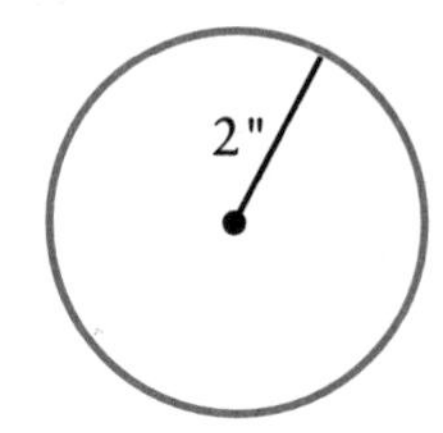

____

7. 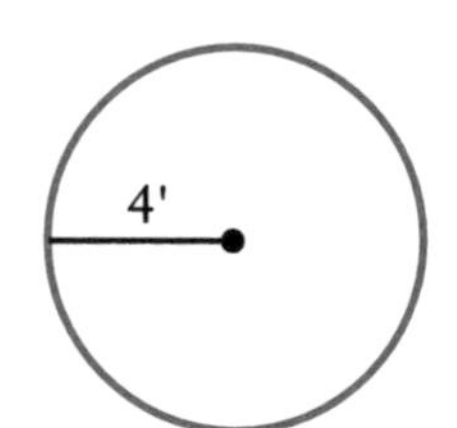

____

8. 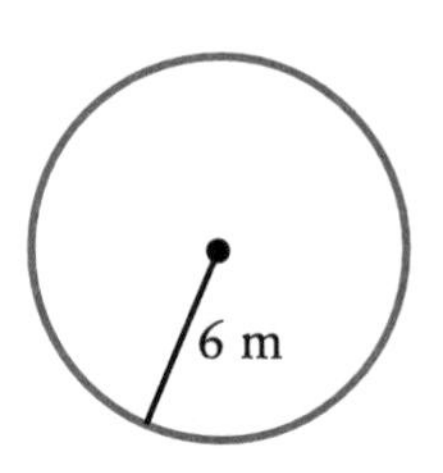

____

9.

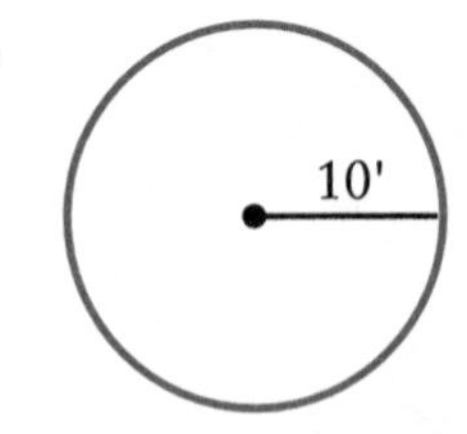

10.

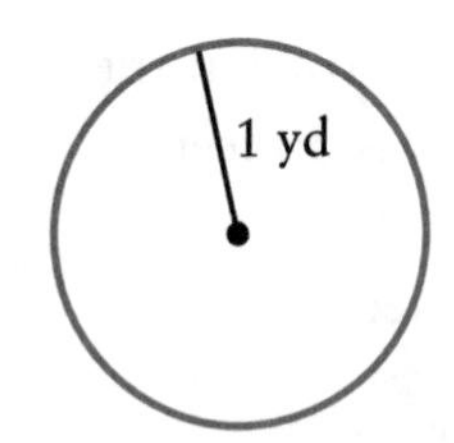

11.

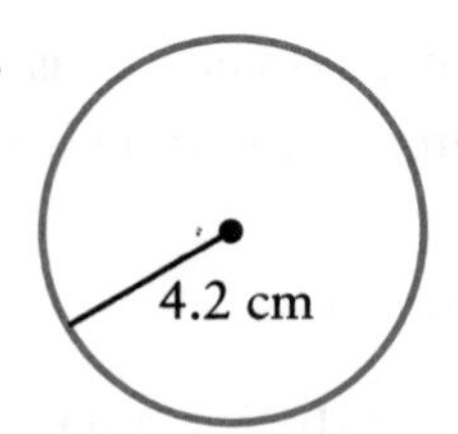

Find the radius of each circle.

12.

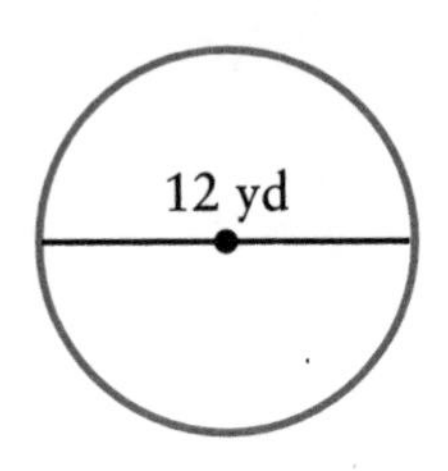

13.

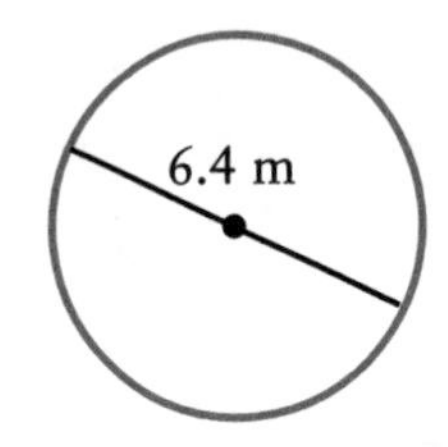

14.

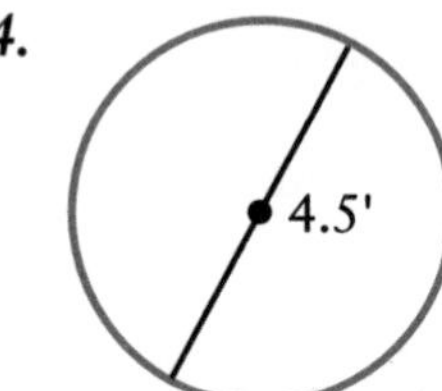

15.

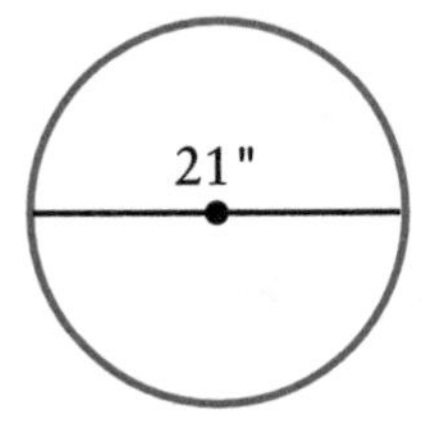

16.

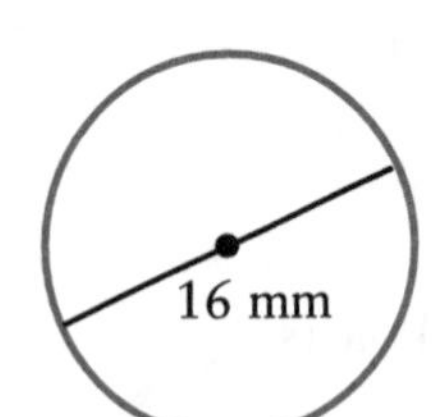

17.

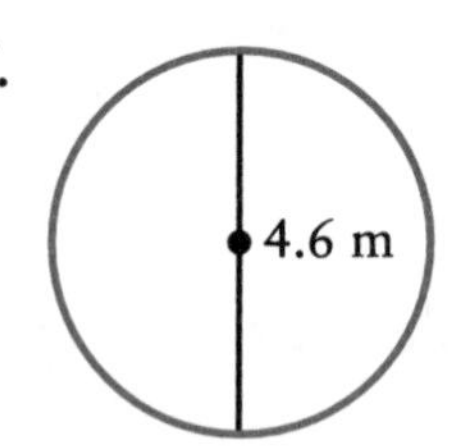

Find the circumference of each circle.

18.

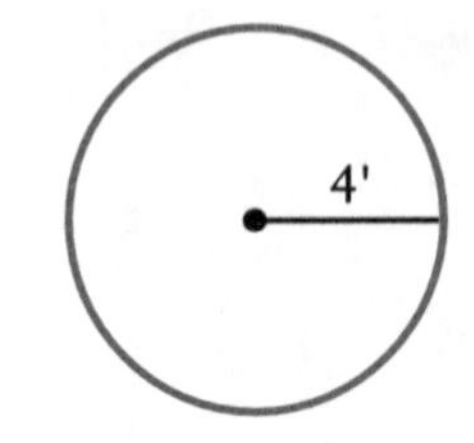

19.

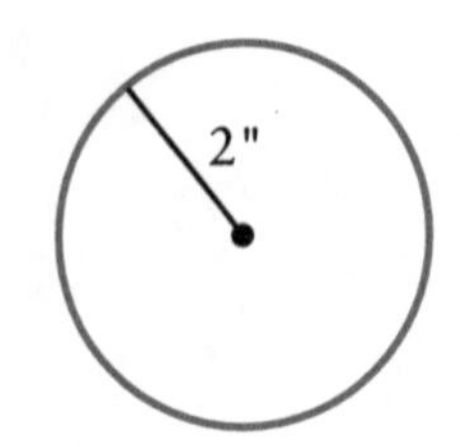

20.

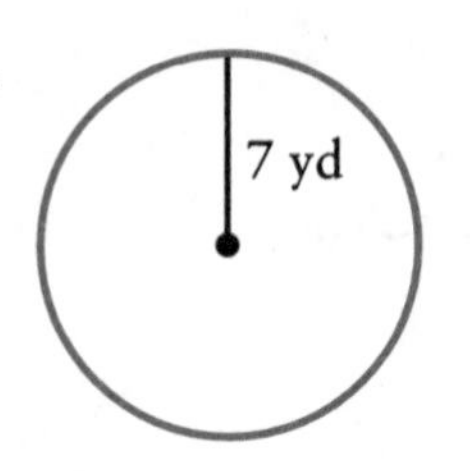

21.

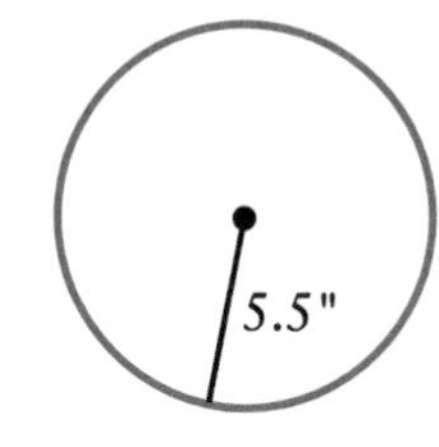

22.

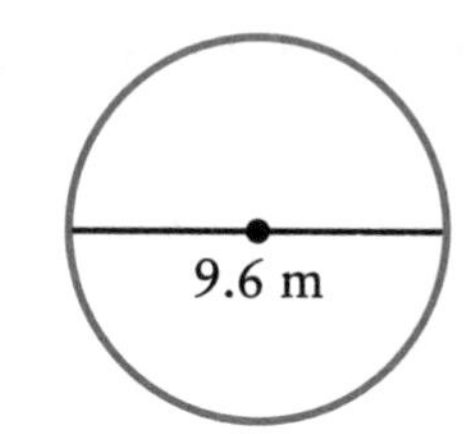

23.

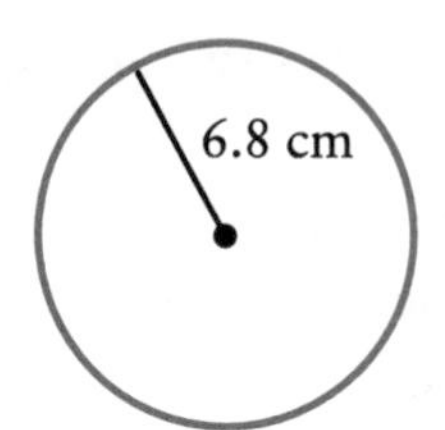

______________ ______________ ______________

24. Charles needs to put trim around a circular flower bed. The diameter of the flower bed is 20 feet. What is the radius of the flower bed? What is the circumference of the flower bed?

25. Hanna wants to add lace trim on the edge of a round table cloth. If the diameter of the table cloth is 48", how much trim does she need? The lace trim is sold in packages of 5'. How many packages will Hanna need to buy?

26. Jon jogs on a circular track. The diameter of the track is 120 yards. What is the circumference? How far will he jog if he does eight laps?

To check your answers, turn to page 123.

Finding Area

Sherelle has a circular swimming pool that she wants to cover when the family is not using it. How many square yards is the area of the pool that Sherelle wants to cover?

The area of a circle is the amount that it takes to cover the circle. To find the area (A) of a circle, you must know the radius (r) of the circle. (If you know the diameter, divide by 2 to find the radius.) The formula for the area of a circle is:

$$A = \pi r^2$$

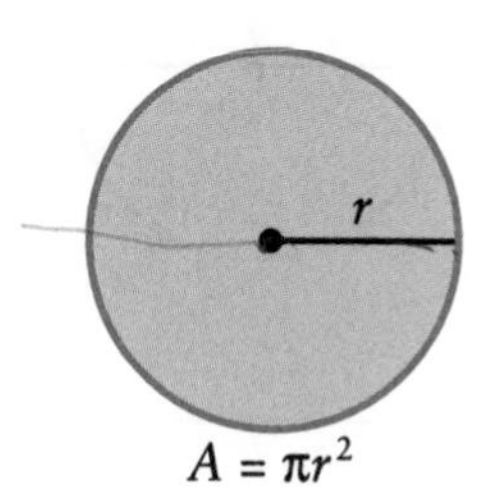

$A = \pi r^2$

Example: Find the area of Sherelle's swimming pool if the radius is 5 yd.

STEP 1. Write the radius of the circle. (If you know the diameter, divide by 2 to find the radius.)

$r = 5$ yards

STEP 2. Substitute the number in the formula. (Use 3.14 for the value of π.)

$A = \pi r^2$

$A = 3.14 \times 5^2$

$A = 3.14 \times 25$

$A = 78.5$ square yards

➡ The cover for Sherelle's swimming pool should be 78.5 square yards.

PRACTICE 23

Fill in the blank.

1. Area is measured in ______________ units.

2. The diameter is ______________ the radius.

3. The ______________ is the distance from the center to the edge of a circle.

4. The formula for the area of a circle is ______________.

Find the radius of each circle.

5.

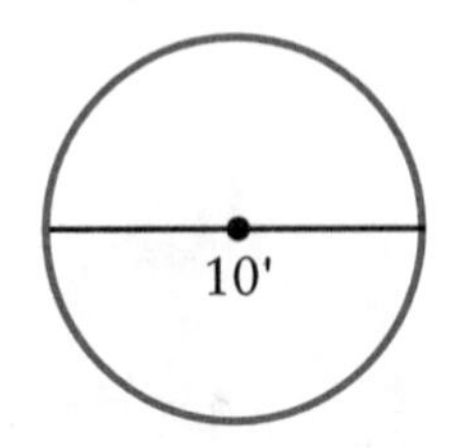

____

6.

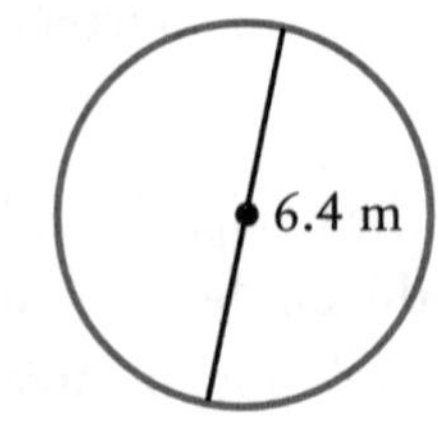

____

7.

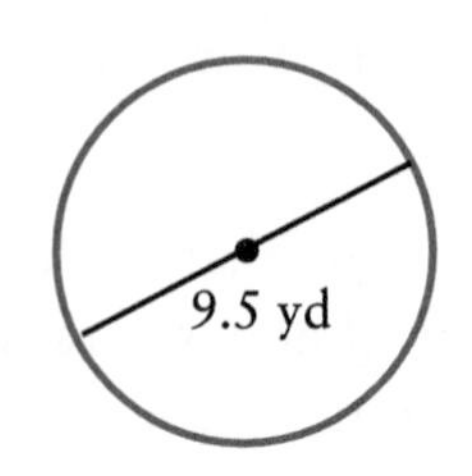

____

Find the area of each circle.

8.

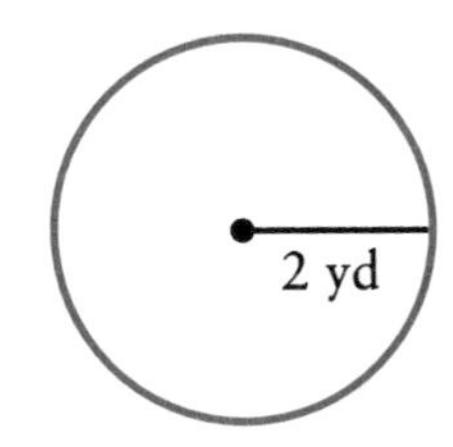

9.

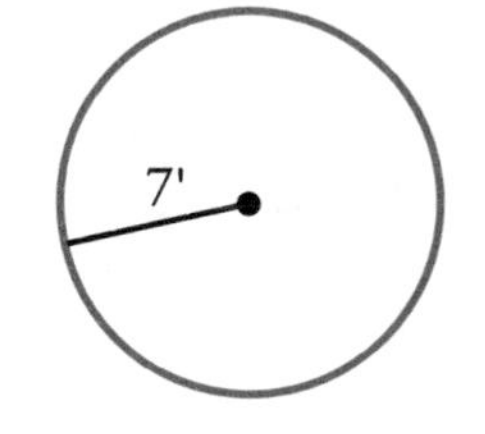

10.

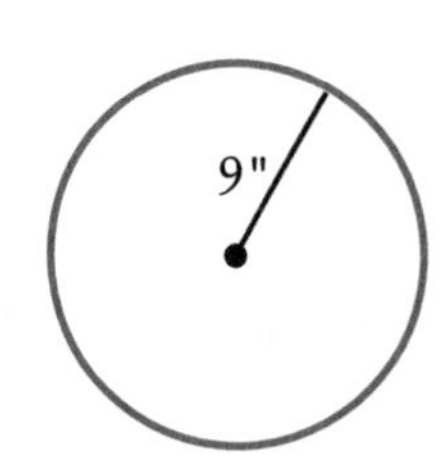

11.

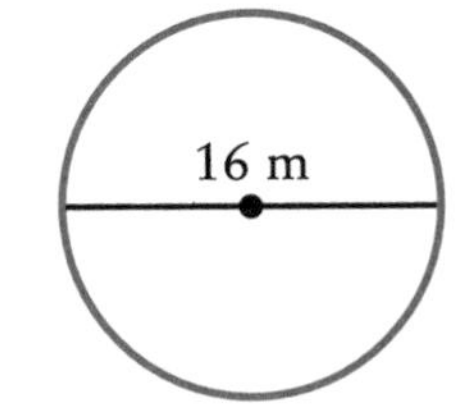

12.

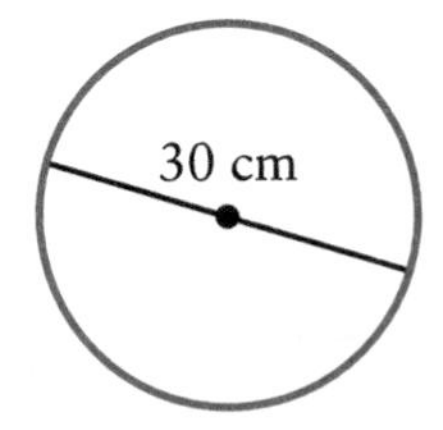

13.

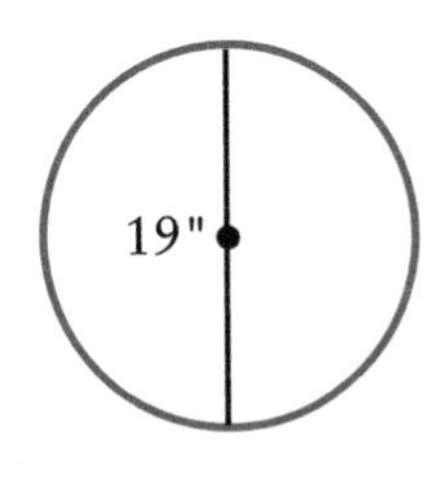

14. Mimi needs to cut a new top for her circular end table. What is the area of the new top if the radius is 12 inches?

15. The Corteras family wants to get a cover for their circular hot tub. The diameter of the tub is 8 feet. What is the area of the cover?

16. Jay is making a circular rug. The diameter of the rug will be 36". What is the radius of the rug? What is the area of the rug?

To check your answers, turn to page 124.

CIRCLES REVIEW

The problems on pages 76 to 78 will help you find out if you need to review the circles section of this book. When you finish, look at the chart on page 78 to see which pages you should review.

Fill in the blanks.

1. The distance around a circle is the _______________.

2. The distance across a circle is the _______________.

3. The _______________ is the distance from the center to the edge of a circle.

4. The formula for the circumference of a circle is _______________.

5. The formula for the area of a circle is _______________.

Find the radius of each.

6.

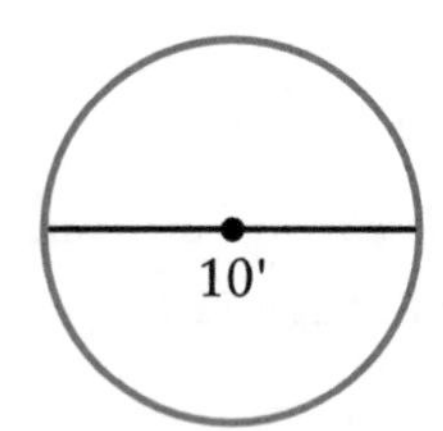

____

7.

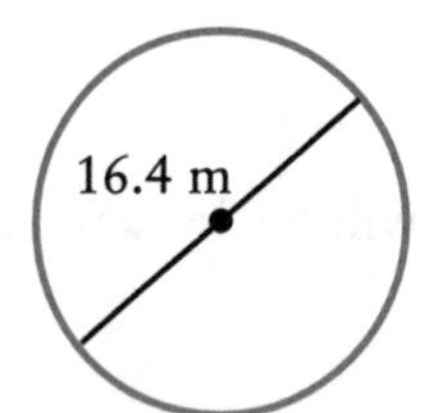

____

8.

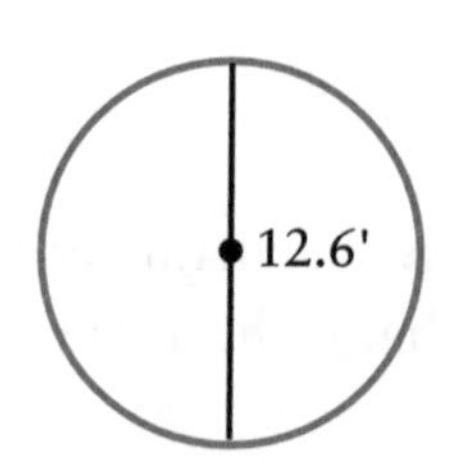

____

Find the diameter of each.

9.

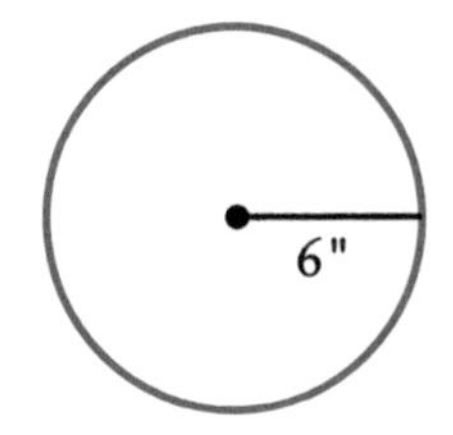

10.

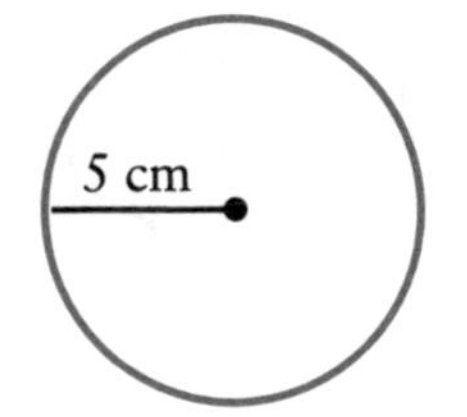

11.

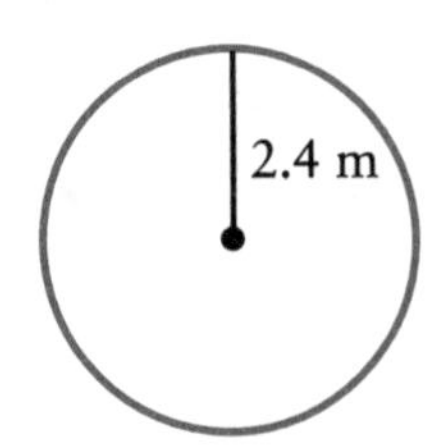

Find the area of each.

12.

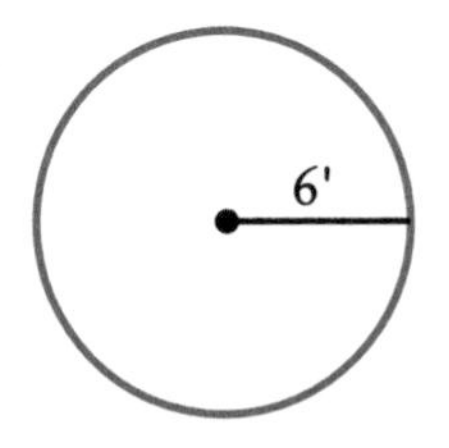

13.

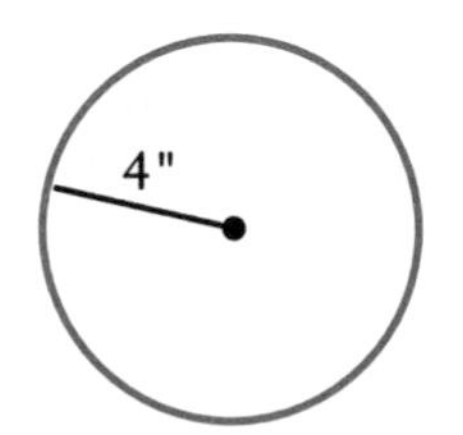

14.

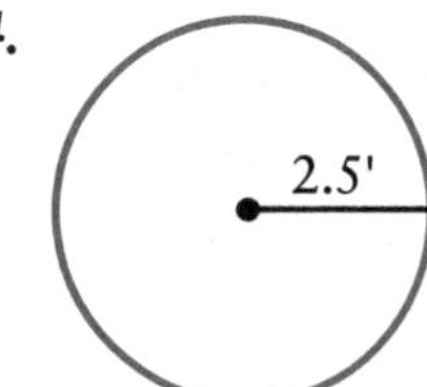

15.

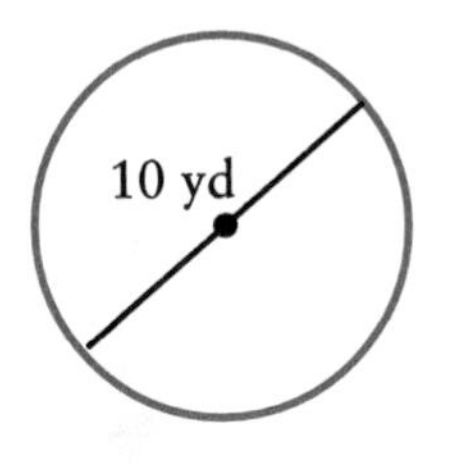

16.

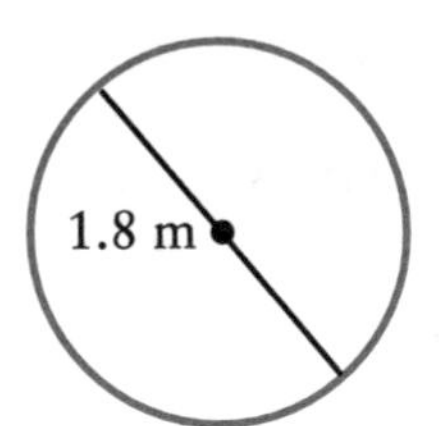

17.

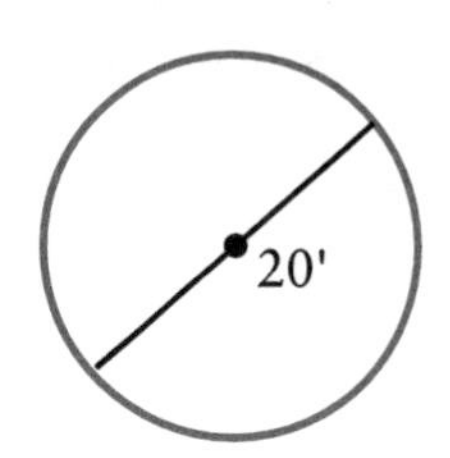

Find the circumference.

18.

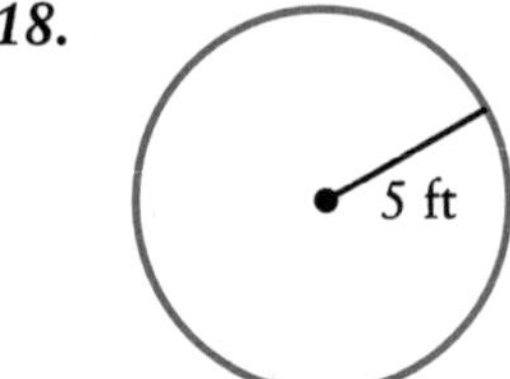

19.

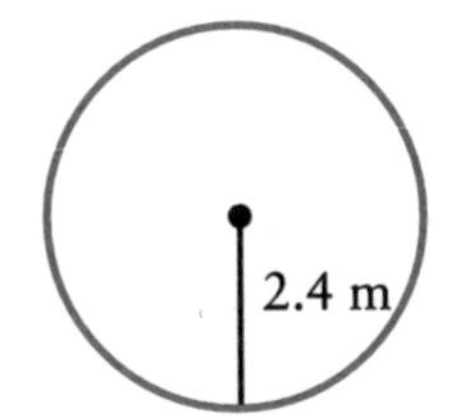

20.

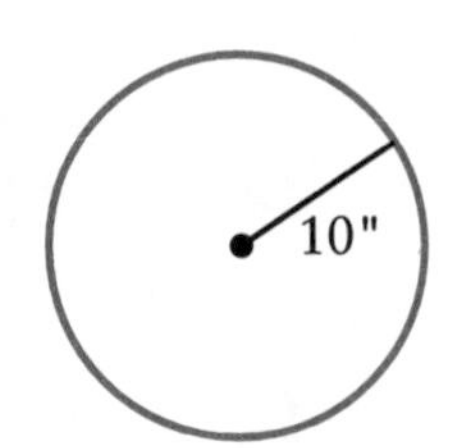

21. 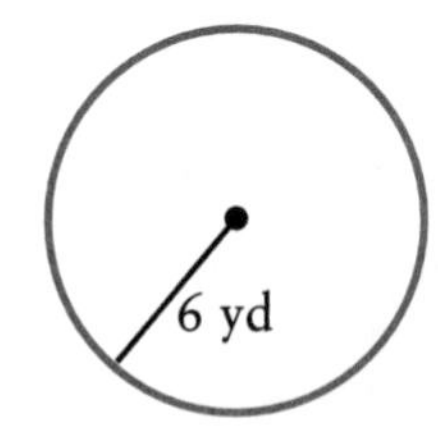

22. 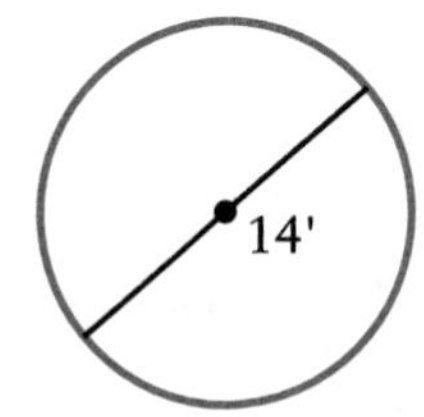

23. 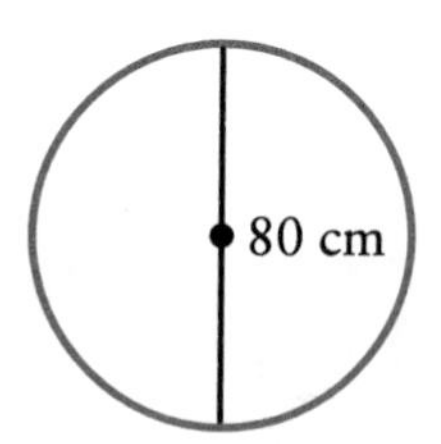

_______________ _______________ _______________

24. The diameter of a circular fountain is 20 feet. What is the circumference of the fountain?

25. A new sealant must be put on the floor of the fountain. One quart of sealant covers 100 sq ft. If the diameter of the fountain floor is 20 feet, how many quarts are needed for the fountain?

PROGRESS CHECK

Check your answers on page 124. Then return to the review pages for the problems you missed. Correct your answers before going on to the next unit.

If you missed problems	*Review pages*
1 to 4, 6 to 11, 18 to 23, or 24	70 to 73
5, 12 to 17, or 25	73 to 75

UNIT

5

Solids

Definitions of Solids

Solid objects have more than two dimensions. Balls, boxes, and many other common objects have three dimensions: length, width, and depth (or height).

There are five different types of solid objects covered in this unit. **Cubes** are three-dimensional objects made of right angles in which all three dimensions are of equal length, width, and height. In a **rectangular solid,** each flat surface forms a rectangle. Unlike the cube, the length, width, and height may be different. **Spheres** are round objects—whichever way you view them, the object looks like a ball. A **cylinder** is a round, tube-like object. Viewed from the top or bottom, it forms a circle. A solid object with a circular base at one end and a point at the other end is called a **cone.**

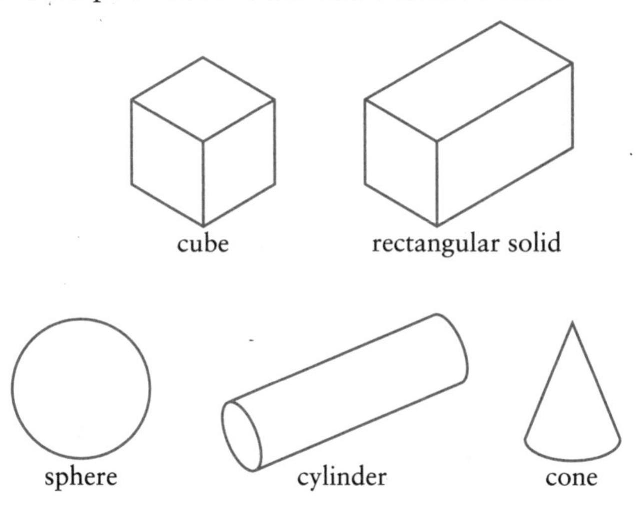

Example: Find the type of solid machine part shown here.

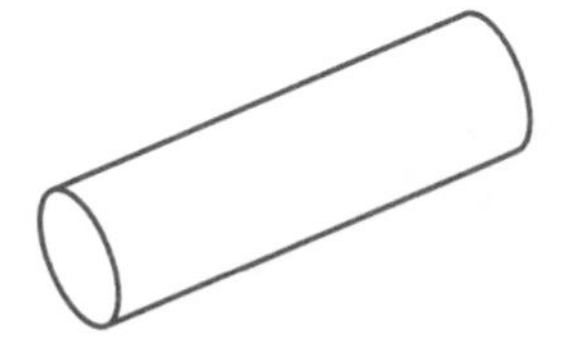

Step 1. Determine the length of the 3 dimensions, the angles, and shape of the ends of the solid.

Circle at both ends, tube connecting the two ends

Step 2. Find the kind of solid shape using the following table:

	Dimensions	*Angles*	*Other*
cube	equal	right	—
rectangular solid	not equal	right	—
sphere	round	none	circle
cylinder	tube	none	circle
cone	circular base	—	point at end

➡ The machine part is in the shape of a cylinder.

PRACTICE 24

Identify each type of solid.

1. 

____

2. 

____

3. 

____

4. 

____

5. 

____

6. 

____

7. 

____

8. 

____

9. 

____

10. 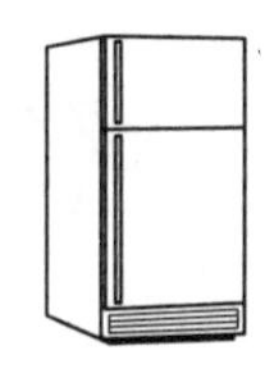

____

11. 

____

12. 

____

To check your answers, turn to page 125.

Definition of Volume

Carlos has a truck which can hold 60 cubic feet of dirt. This is the measure of the volume of the truck.

Volume is a measurement of the amount of space within a solid object. Volume is measured in cubic units, such as cubic inches, cubic feet, and cubic centimeters. These measurements can also be shown as in^3, ft^3, or cm^3.

To find the volume of a solid object, you must know the dimensions of the object. In the case of a cube, you only have to know the length of one of the sides, since all of the sides of a cube are equal in length. For rectangular solids, you need to measure the height, width, and length of the object. You can find the volume of a cylinder once you know the radius of the circle at either end and the length or height of the tube.

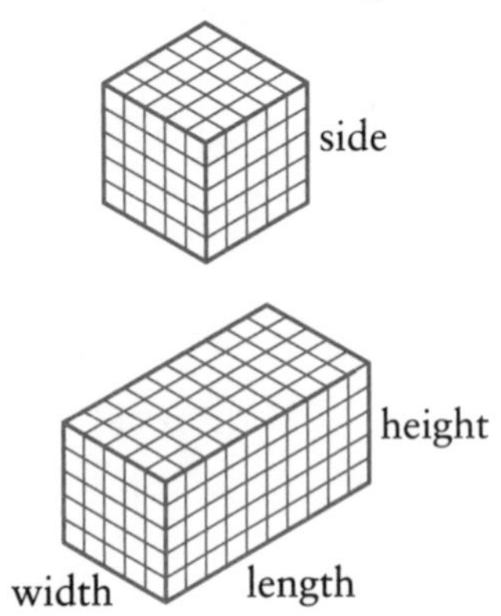

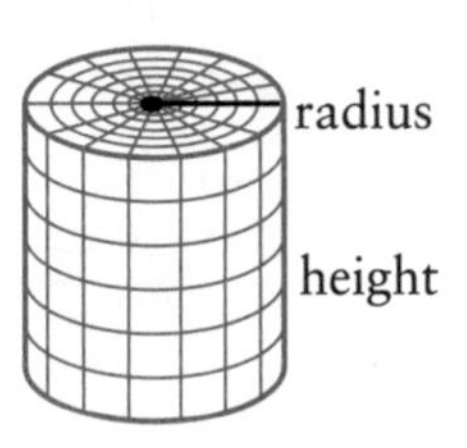

PRACTICE 25

Fill in the blanks.

1. The amount of space within an object is the ______________.

2. An object must be ______________-dimensional to have volume.

3. Volume is measured in ______________ units.

4. Suggest three times you might measure volume in your home.

5. Suggest three times you might measure volume in your community.

Write *yes* if you can find the volume of the object; *no* if you can not.

6. 

____

7. 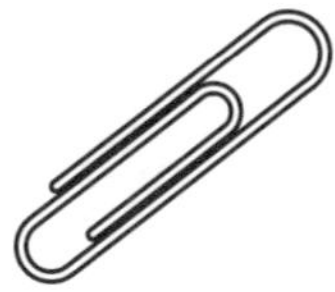

____

8. 

____

9. 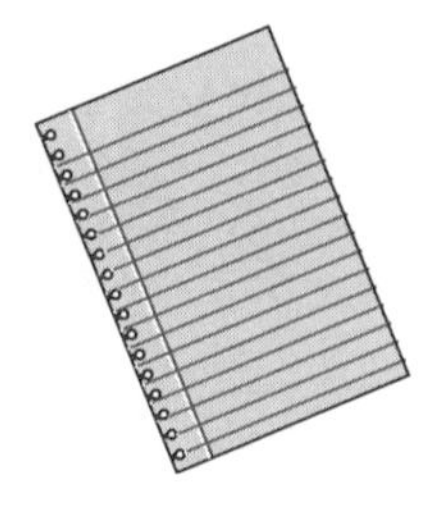

____

10. 

____

11. 

____

12. 

____

13. 

____

14. 

____

To check your answers, turn to page 125.

Volume of a Cube

Gustavo has a box which looks like a square from all sides. One edge (or side) is 3 ft long. How much can the box hold?

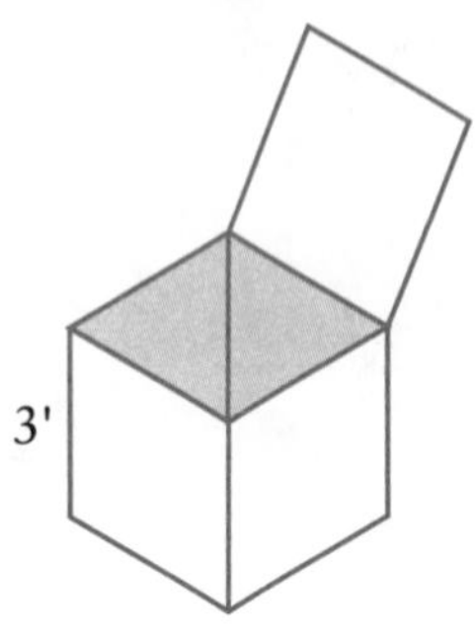

A box which forms a square on each of its six sides is a cube. A cube is a special kind of rectangular solid. To find the volume (V) of a cube, you need to measure one side (s) and use the formula

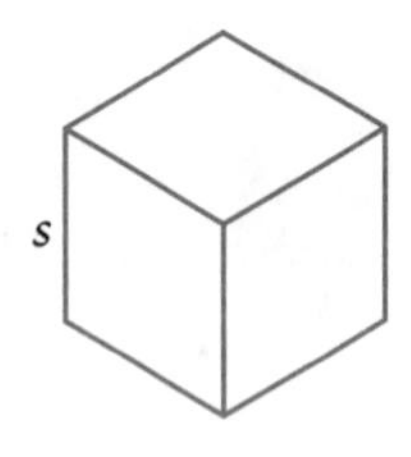

$$V = s^3 \quad \text{or} \quad V = s \times s \times s.$$

Example: Find the volume of Gustavo's box.

Step 1. Write the length of any side. $s = 3$ feet

Step 2. Substitute the number into the formula for the volume of a cube. $V = s^3$
Solve the formula.
$V = 3^3$
$V = 3 \times 3 \times 3$
$V = 27$ cubic feet

➡ The volume of Gustavo's box is 27 cubic feet or 27 ft^3.

PRACTICE 26

Find the volume of each cube.

1.

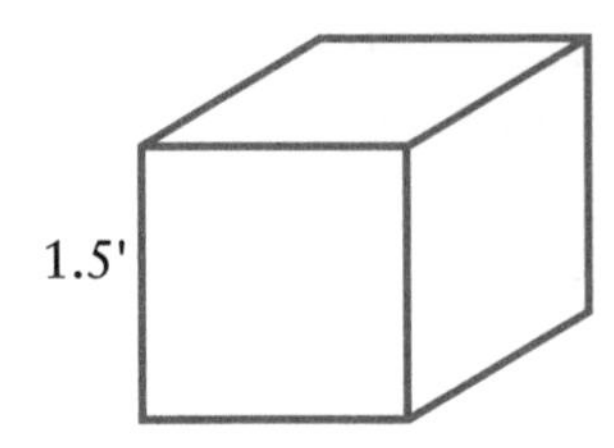

$s = 1.5'$

$V =$ ________

2.

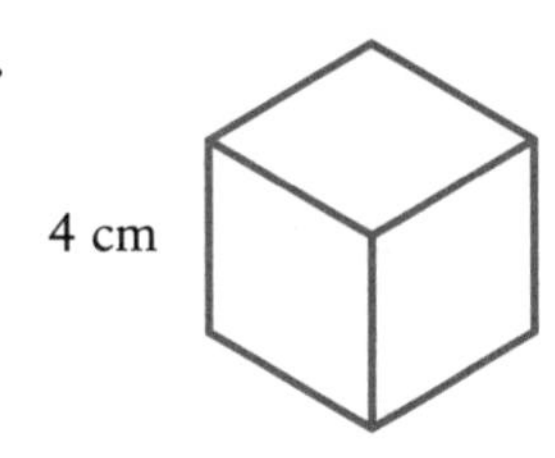

$s = 4$ cm

$V =$ ________

3.

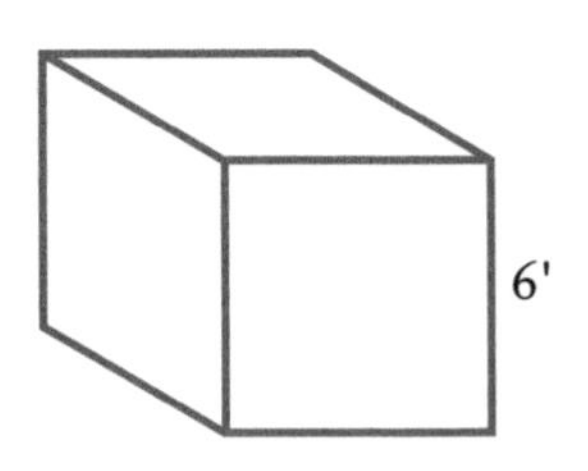

$s = 6'$

$V =$ ________

4.

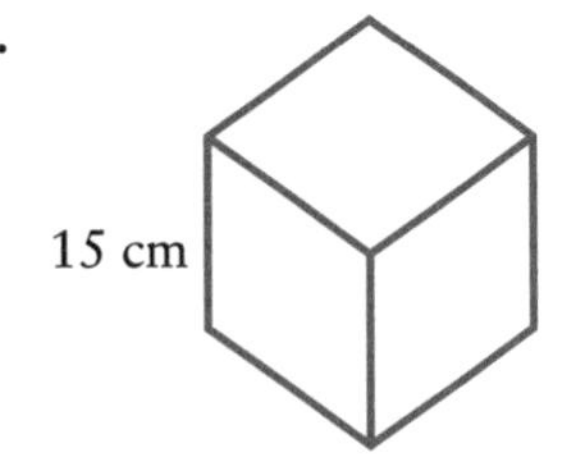

$s = 15$ cm

$V =$ ________

5.

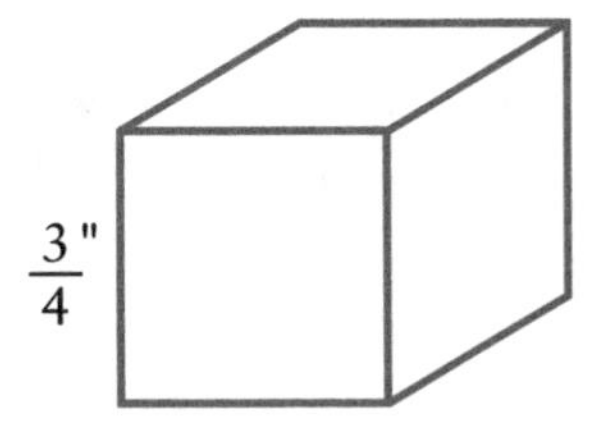

$s = \frac{3}{4}''$

$V =$ ________

6.

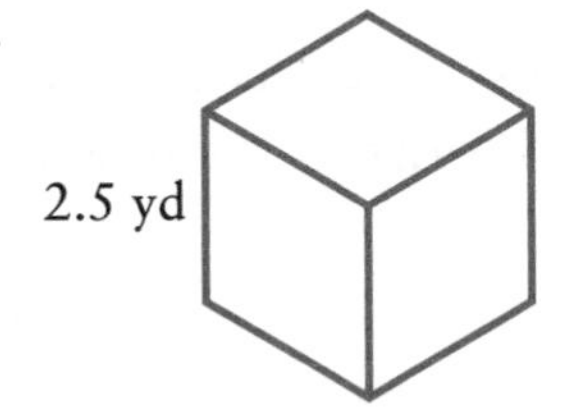

$s = 2.5$ yd

$V =$ ________

7.

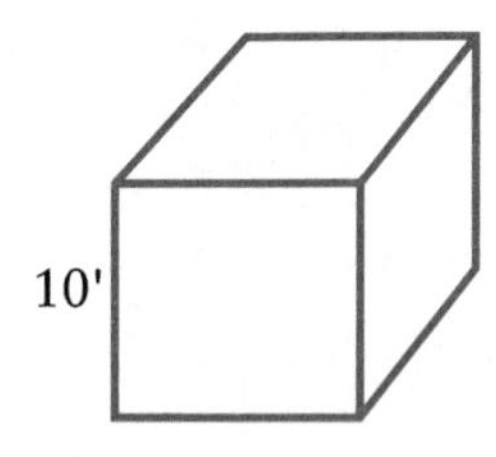

$s = 10'$

$V =$ ________

8.

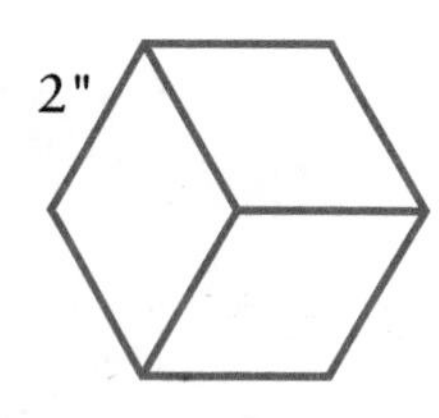

$s = 2"$

$V =$ ________

9.

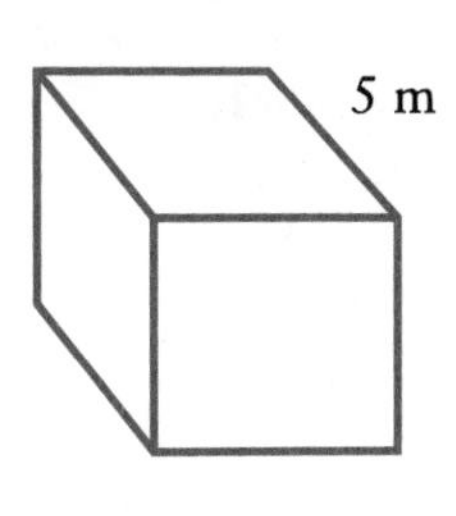

$s = 5$ m

$V =$ ________

10. Berta has a small gift box that is shaped like a cube. The cube is 3 inches on a side. What is the volume of the box?

11. Tim is working for a company that installs underground sprinkler systems. He must dig a hole that has 4.5' sides. If the hole is a cube, how much dirt will he move?

To check your answers, turn to page 125.

Volume of a Rectangular Solid

A solid object that looks like a rectangle from each side is called a rectangular solid. To find the volume (V) of a rectangular solid, multiply the measurements for the length (l), width (w), and height (h) of the solid. The formula is $V = lwh$.

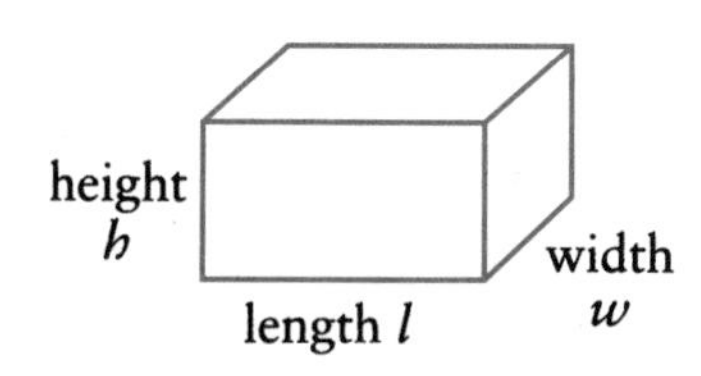

9 in.

10 in.

18 in.

Example: Find the amount of water this fish tank can hold.

STEP 1. Write the measurements for each dimension of the object.

$l = 18$ inches
$h = 9$ inches
$w = 10$ inches

STEP 2. Use the formula for the volume of a rectangular solid. Substitute and solve.

$V = lwh$
$V = 18 \times 10 \times 9$
$V = 180 \times 9$
$V = 1{,}620$ cubic inches

➠ The fish tank can hold 1,620 cubic inches of water.

PRACTICE 27

A. Find the volume of each.

1.

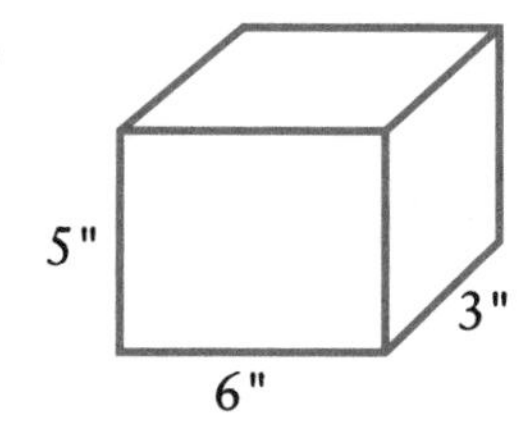

$l = 6"$
$w = 3"$
$h = 5"$

$V =$ ________

2.

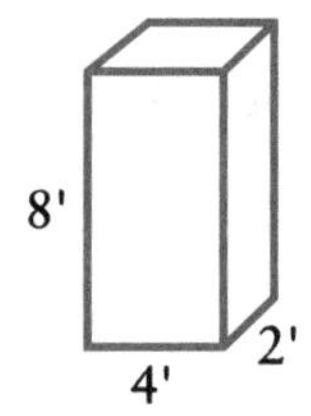

$l = 4'$
$w = 2'$
$h = 8'$

$V =$ ________

3.

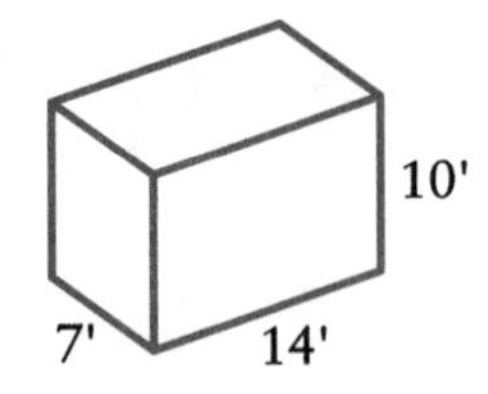

$l = 14'$
$w = 7'$
$h = 10'$

V = ________

4.

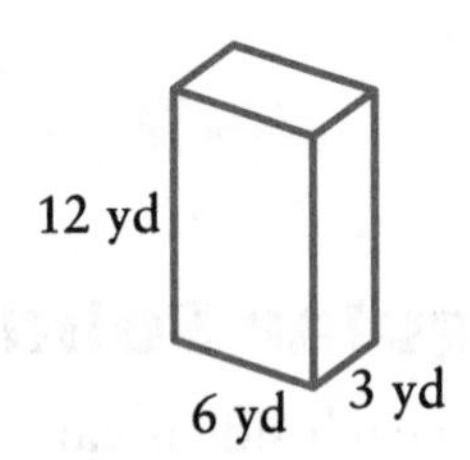

$l = 6$ yd
$w = 3$ yd
$h = 12$ yd

V = ________

5.

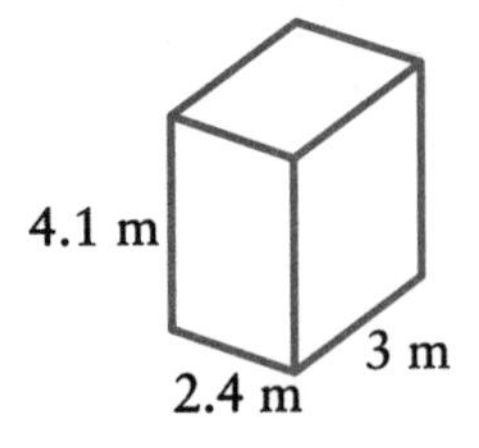

$l = 3$ m
$w = 2.4$ m
$h = 4.1$ m

V = ________

6.

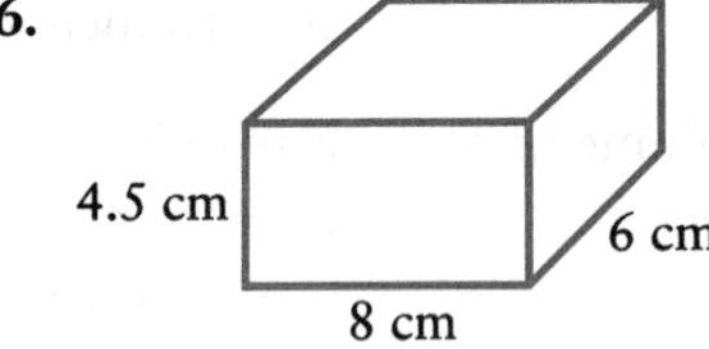

$l = 8$ cm
$w = 6$ cm
$h = 4.5$ cm

V = ________

To find the volume of a rectangular solid, all three dimensions must be stated in the same units. Sometimes you must convert a dimension to find the volume.

Example: Ralph measured a box as 14" long, 1 foot wide, and 15" high. What is the volume of the box?

Step 1.	Convert the dimension.	1 foot = 12 in
Step 2.	Substitute the numbers in the formula and solve.	$V = lwh$ $V = 14 \times 12 \times 15$ $V = 2{,}520$ cu in

➡ The volume of the box is 2,520 cu in.

B. Find the volume of each.

7.

$l = 10"$
$w = 6"$
$h = 1\frac{1}{2}'$

$V =$ ________

8.

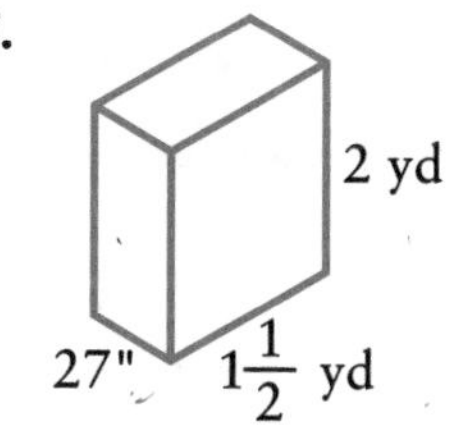

$l = 1\frac{1}{2}$ yd
$w = 27"$
$h = 2$ yd

$V =$ ________

9.

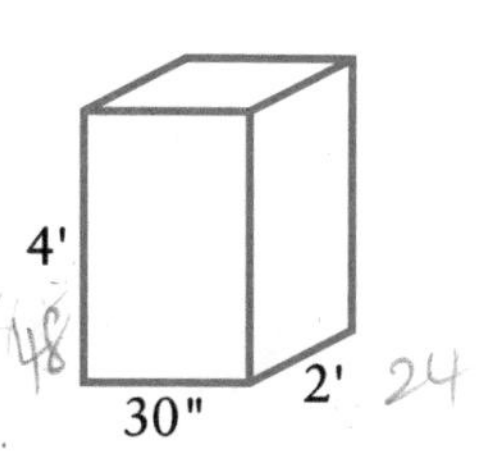

$l = 30"$
$w = 2'$
$h = 4'$

$V =$ ________

10.

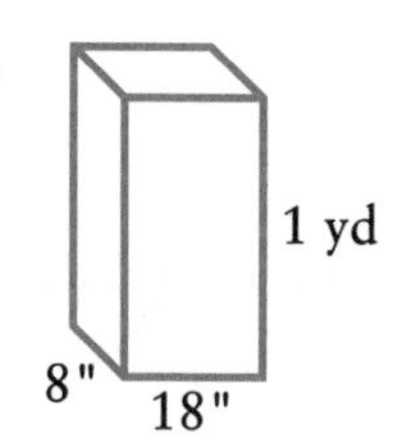

$l = 18"$
$w = 8"$
$h = 1$ yd

$V =$ ________

11.

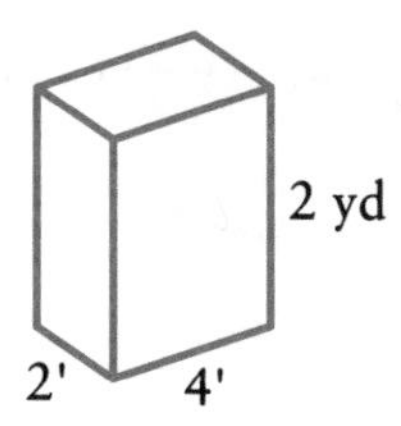

$l = 4'$
$w = 2'$
$h = 2$ yd

$V =$ ________

12.

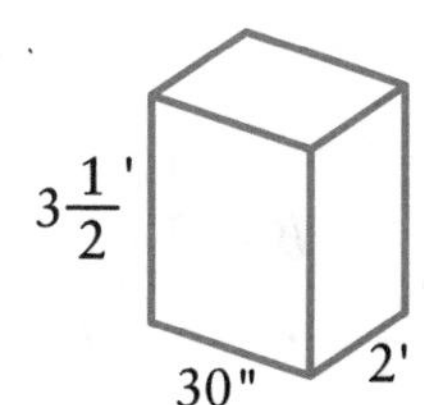

$l = 30"$
$w = 2'$
$h = 3\frac{1}{2}'$

$V =$ ________

13. What is the volume of a shoebox 1' by 4 inches by 8"?

14. What is the volume of dirt removed from a hole that measures 3 yd by 12 feet by 5 yds?

15. A cereal box measures 3" by 8" by 11". What is the volume of the cereal box?

To check your answers, turn to page 126.

Volume of a Cylinder

Jung has an outdoor swimming pool to which he needs to add chlorine. The directions tell him how much chlorine to add per cubic foot of water. What is the volume of Jung's pool?

A cylinder is a solid object with circles at either end. To find the volume (V) of a cylinder, find the radius (r) of the circle at either end and the height (h). The height of the cylinder is the distance from one circular end to the other.

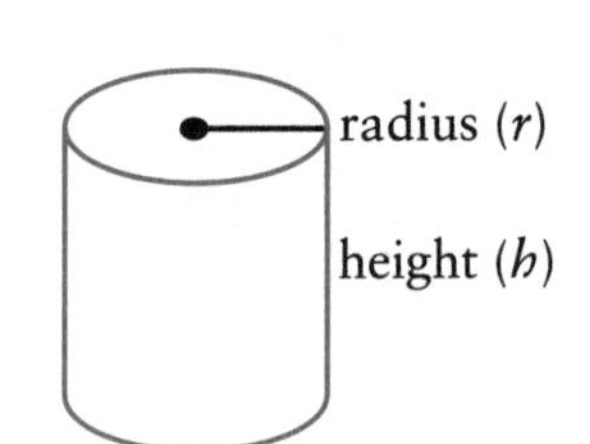

The formula for the volume of a cylinder is $V = \pi r^2 h$.

Example: Find the volume of Jung's pool if the radius is 4 yd and the height is 2 yd.

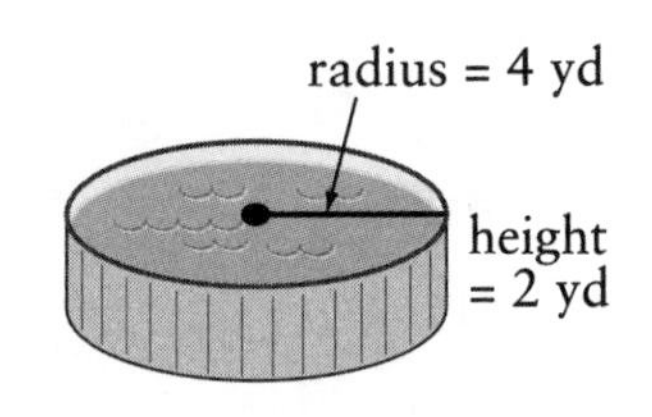

STEP 1.	Write the measurements you will need.	$r = 4$ yd $h = 2$ yd
STEP 2.	Use the formula for the volume of a cylinder.	$V = \pi r^2 h$
STEP 3.	Substitute and solve the formula.	$V = \pi \times r \times r \times h$ $V = 3.14 \times 4 \times 4 \times 2$ $V = 100.48$ cubic yards

Remember to use 3.14 for pi.

➡ The volume of Jung's swimming pool, rounded to the nearest cubic yard, is 100 cubic yards.

PRACTICE 28

Find the volume of each cylinder.

1.

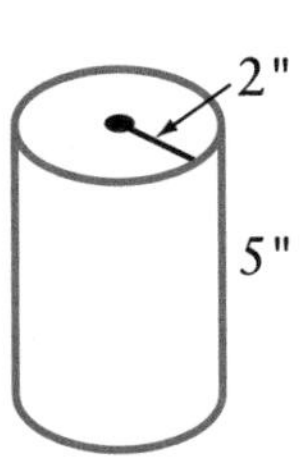

$r = 2"$
$h = 5"$

$V =$ ________

2.

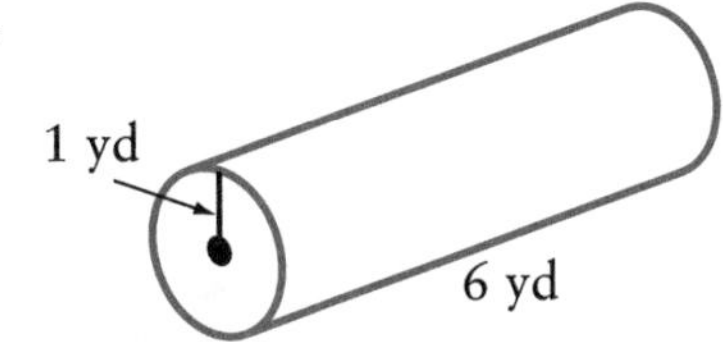

$r = 1$ yd
$h = 6$ yd

$V =$ ________

3.

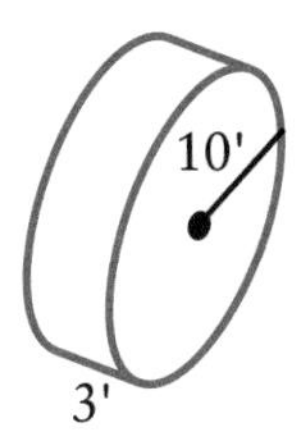

$r = 10'$
$h = 3'$

$V =$ ________

4.

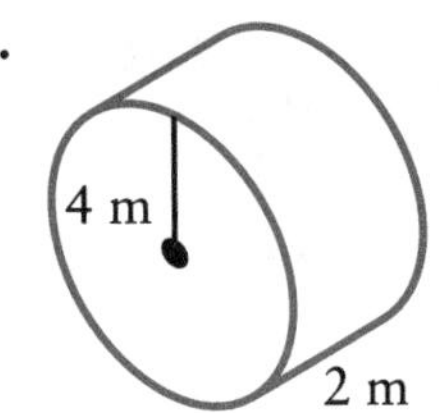

$r = 4$ m
$h = 2$ m

$V =$ ________

5.

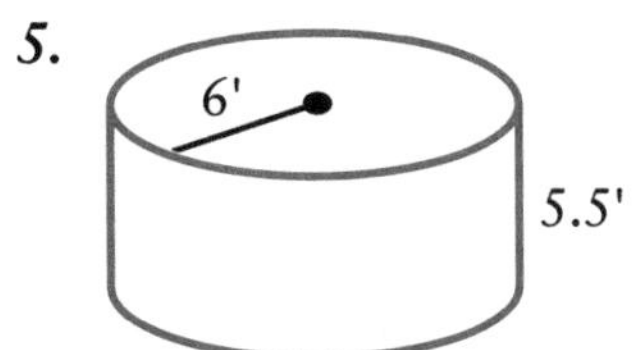

$r = 6'$
$h = 5.5'$

$V =$ ________

6.

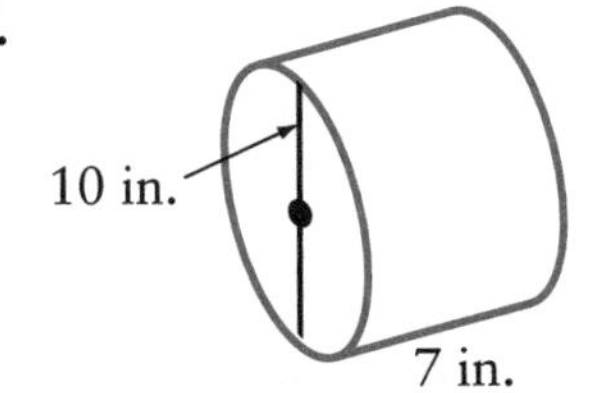

$d = 10$ in
$h = 7$ in

$V =$ ________

Find the volume of each cylinder. Make sure both dimensions are stated in the same units of measure before you use the volume formula.

7.

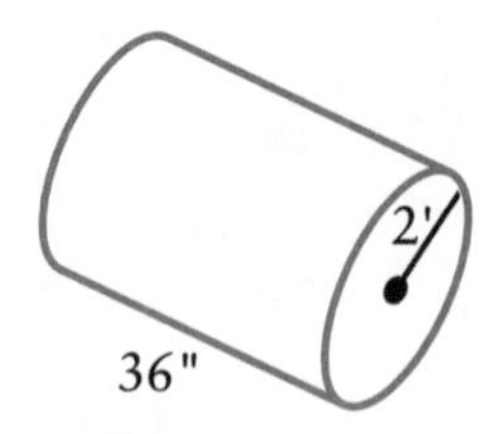

$r = 2'$
$h = 36"$

$V =$ ________

8.

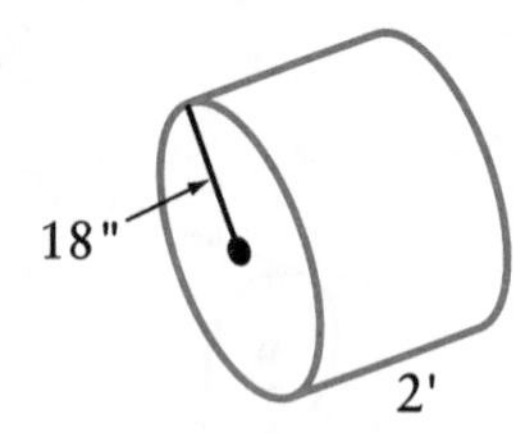

$r = 18"$
$h = 2'$

$V =$ ________

9.

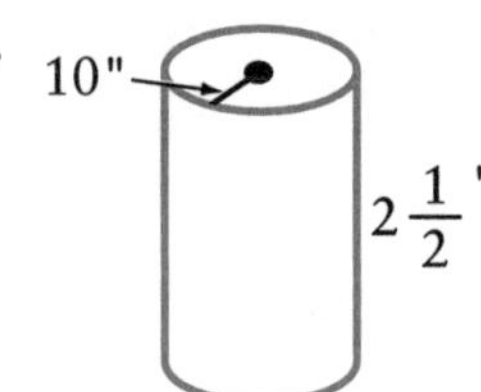

$r = 10"$
$h = 2\frac{1}{2}'$

$V =$ ________

10.

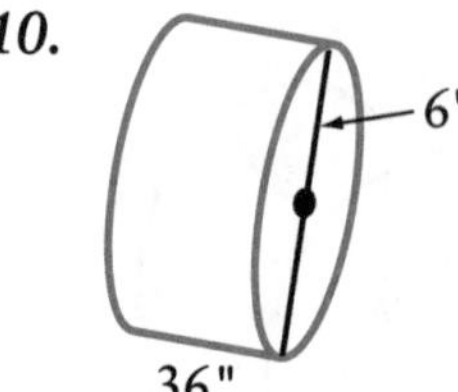

$d = 6'$
$h = 36"$

$V =$ ________

11. The height of a grain bin is 18'. The radius of the bin is 10'. What is the volume of the grain bin?

12. Su Yong works in a mechanic shop. Oil is purchased in drums, which are really just large cylinders. The height of the cylinder is 4'. The radius is 2'. What is the volume of the drum?

13. The weight of a cubic foot of oil is 5 lbs. How much does the oil in the drum in item 12 weigh?

To check your answers, turn to page 126.

SOLIDS REVIEW

The problems on pages 93 to 95 will help you find out if you need to review the solids section of this book. When you finish, look at the chart on page 95 to see which pages you should review.

Name the type of solid.

1.

____

2.

____

3.

____

4.

____

Match the volume formulas to the solids.

______ **5.** cube — *a.* $V = \pi r^2 h$

______ **6.** rectangular solid — *b.* $V = lwh$

______ **7.** cylinder — *c.* $V = s^3$

Find the volume.

8.

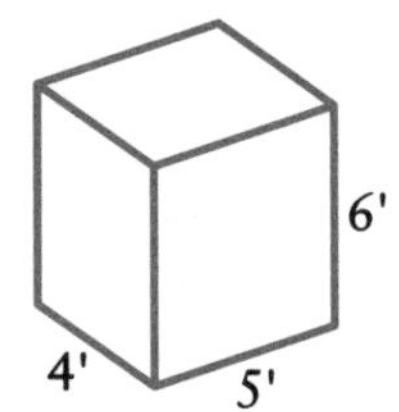

$h = 6'$
$w = 4'$
$l = 5'$

$V =$ ________

9.

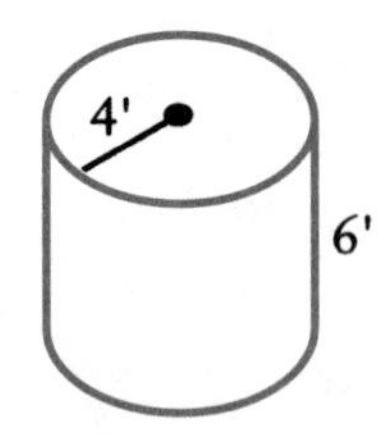

$r = 4'$
$h = 6'$

$V =$ ________

10.

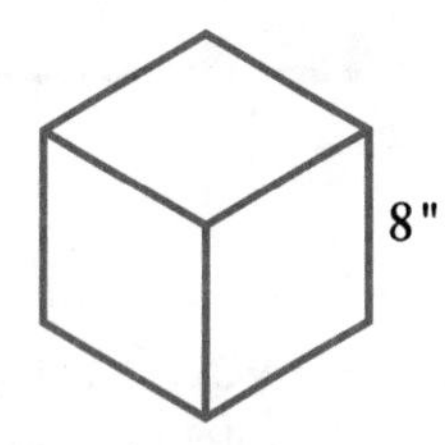

$s = 8''$

$V =$ ________

11.

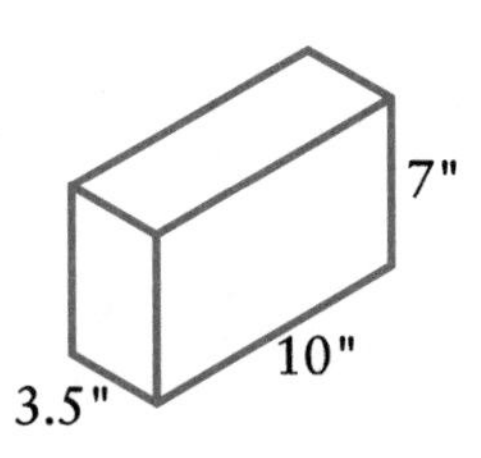

$h = 7''$
$w = 3.5''$
$l = 10''$

$V =$ ________

12.

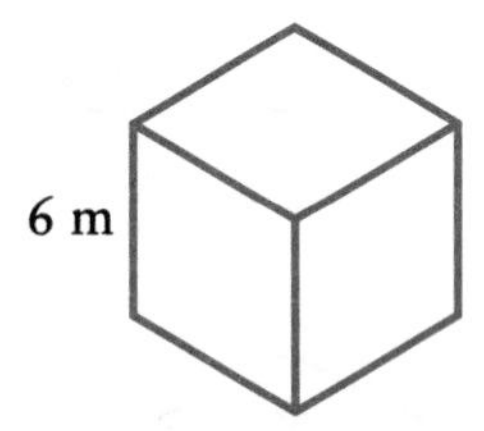

$s = 6$ m

$V =$ ________

13.

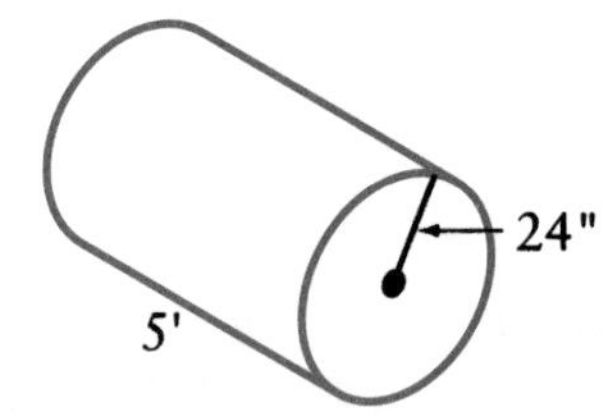

$r = 24''$
$h = 5'$

$V =$ ________

14.

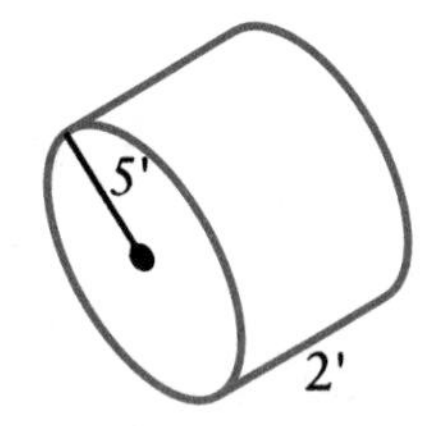

$r = 5'$
$h = 2'$

$V =$ ________

15.

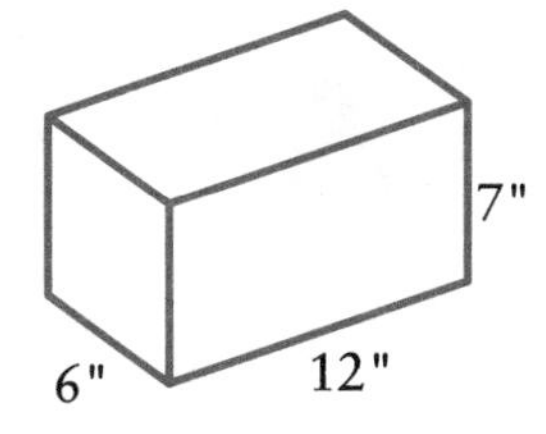

$l = 12"$
$w = 6"$
$h = 7"$

$V =$ ________

16.

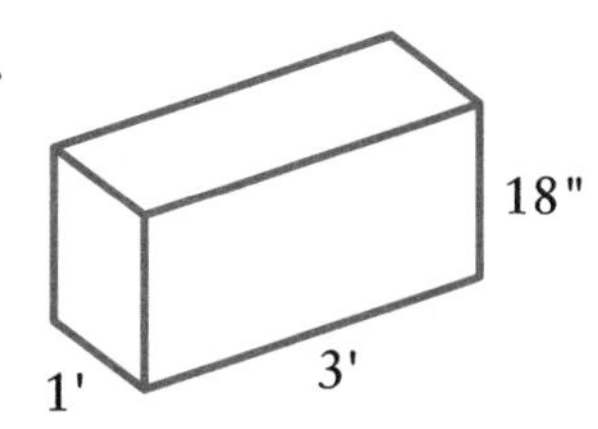

$h = 18"$
$w = 1'$
$l = 3'$

$V =$ ________

17. The inside dimensions of a refrigerator are $2\frac{1}{4}'$ by 3' by 4'. What is the volume of the refrigerator?

18. A cylindrical oil drum is 2 meters high and has a 1.5 meter radius. How many cubic meters of oil can the drum contain?

19. Sally needs to order concrete for her driveway, which is 12 yards long and 5 yards wide. The concrete will be 9" thick. How many cubic yards of concrete must she order?

PROGRESS CHECK

Check your answers on page 127. Then return to the review pages for the problems you missed. Correct your answers before going on to the next unit.

If you missed problems	*Review pages*
1 to 4	79 to 81
5, 8, 10, or 12	84 to 86
6, 11, 15–17, 19	87 to 90
7, 9, 13, 14, 18	90 to 92

UNIT

6

The Coordinate Plane

Ordered Pairs

A **coordinate plane** is a graph onto which points can be placed. The plane has two lines called **axes.** The axes are like number lines with marks for each whole number or **integer.** Where the axes meet is marked as 0. The ***x*-axis** runs across the graph from left to right. To the right of the 0, the numbers are marked as 1, 2, 3, and so on. To the left of the 0, the numbers are marked −1, −2, −3, and so on. The ***y*-axis** runs from the top to the bottom of the graph. Positive integers are marked above the 0; negative numbers below the 0.

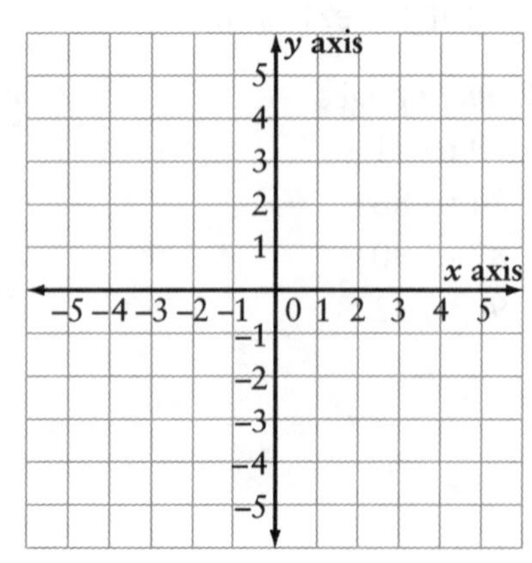

A point is placed on the coordinate plane as an **ordered pair** that tells its position. The ordered pair is made up of two numbers. The first number is the **x-coordinate;** the x-coordinate tells how far the point is from the y-axis. Positive numbers mean that the point is to the right of the y-axis; negative numbers mean the point is to the left.

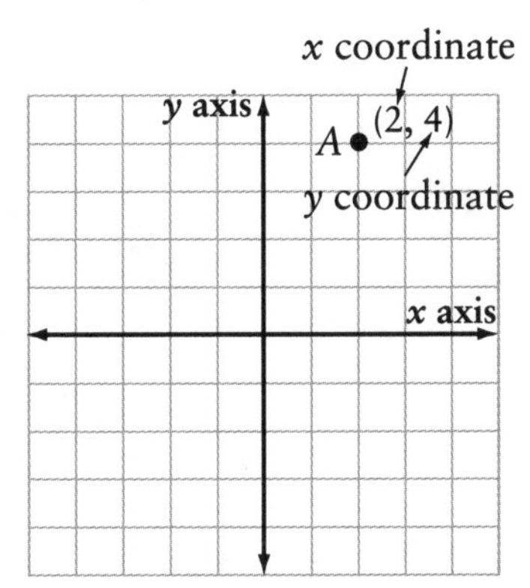

The second number in the ordered pair is the **y-coordinate.** It tells how far the point is from the x-axis. Positive numbers mean the point is above the x-axis; negative numbers mean the point is below the x-axis.

Example: Name the ordered pair that defines the position of point A.

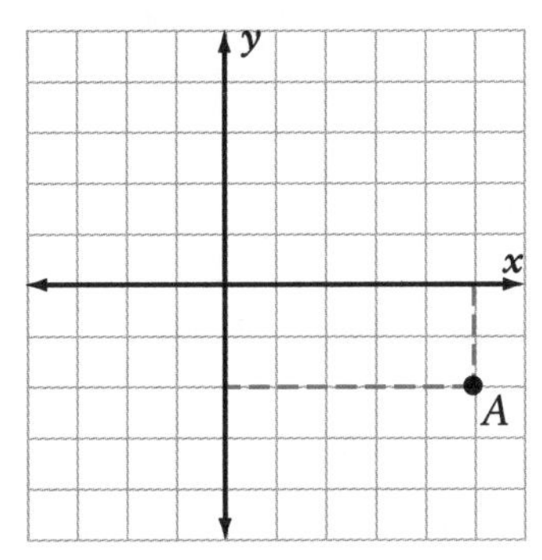

STEP 1.	Find the position of the point on the x-axis.	$x = 5$
STEP 2.	Find the position of the point on the y-axis.	$y = -2$
STEP 3.	Write the ordered pair within parentheses, listing the x-axis position first.	$A = (5, -2)$

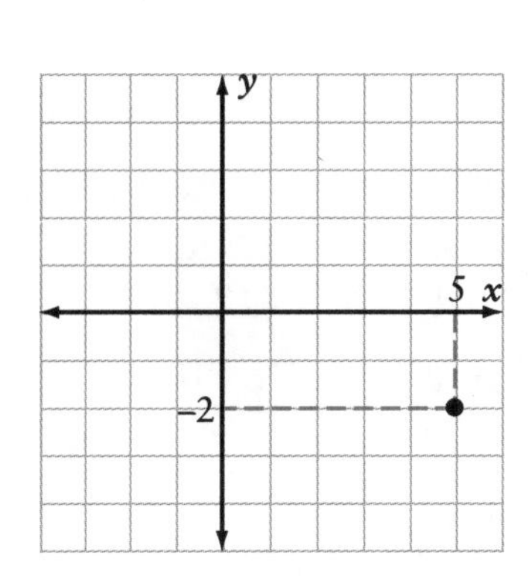

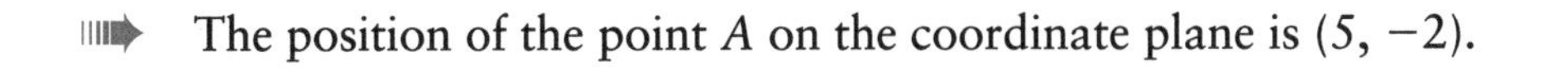

The position of the point A on the coordinate plane is $(5, -2)$.

PRACTICE 29

What is the value of the *x*- and *y*-coordinates for each pair?

1. $(-2, +3)$

$x =$ ________

$y =$ ________

2. $(0, 4)$

$x =$ ________

$y =$ ________

3. $(-3, -2)$

$x =$ ________

$y =$ ________

4. $(2, -4)$

$x =$ ________

$y =$ ________

5. $(12, 8)$

$x =$ ________

$y =$ ________

6. $(3, -2)$

$x =$ ________

$y =$ ________

Use the graph to answer 7–20.

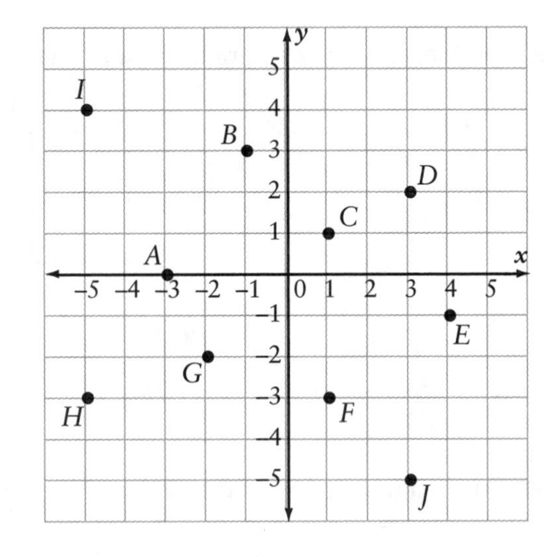

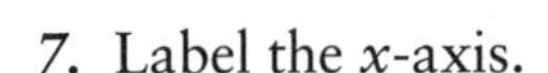

7. Label the *x*-axis.

8. Label the *y*-axis.

9. What is the *x*-coordinate for point *A*?

10. What is the *y*-coordinate for point *A*?

Write the ordered pair for each set of coordinates.

11. *A* **12.** *B* **13.** *C* **14.** *D* **15.** *E*

16. *F* **17.** *G* **18.** *H* **19.** *I* **20.** *J*

To check your answers, turn to page 127.

Plotting Pairs on a Coordinate Graph

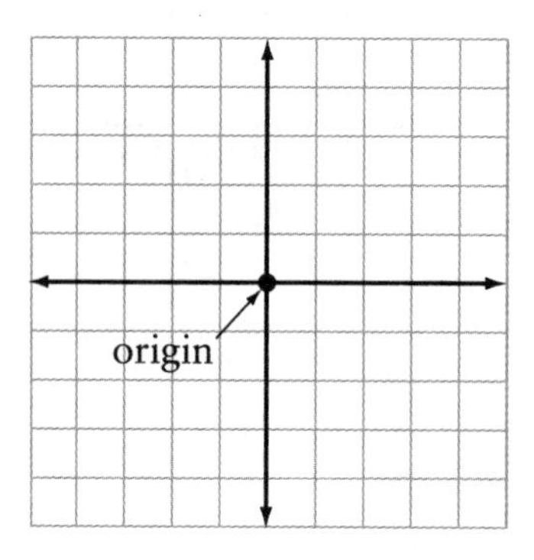

To **plot** a point on a coordinate graph, you must find the x- and y-coordinates of the point. The center of the graph is called the **origin.** The origin has an x-coordinate of 0 and a y-coordinate of 0.

When plotting a point, find the position along the x axis by moving from the origin to the right for a positive number or to the left for a negative number. You move as many spaces as indicated by the first number in the ordered pair. For example, if the x-coordinate is 7, move seven spaces to the right from the origin. You then find the position on the y-axis by moving up for a positive number and down for a negative number. Once again, you move as many spaces as indicated by the second number in the ordered pair. Finally, you draw a point and label it.

Example: Plot point B at $(-3, 2)$.

Step 1.	Begin at the origin.	Origin = (0, 0)
Step 2.	Move the number of spaces along the x-axis indicated by the x-coordinate.	$x = -3$
Step 3.	Move the number of spaces along the y-axis indicated by the y-coordinate.	$y = 2$
Step 4.	Draw the point and label it.	

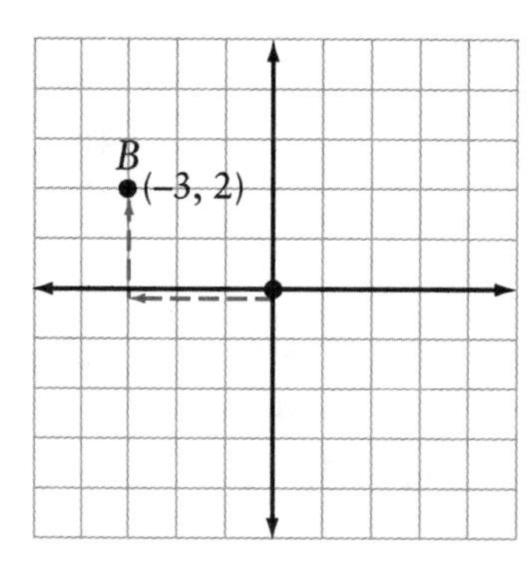

➡ Point B is labeled at $(-3, 2)$ on the grid.

PRACTICE 30

Fill in the blank.

1. The x coordinate indicates how far ____________ or ____________ to move from the origin.

2. The y coordinate indicates how far ____________ or ____________ to move from the origin.

3. The coordinates for the origin are ____________.

Plot the points on each graph.

4.

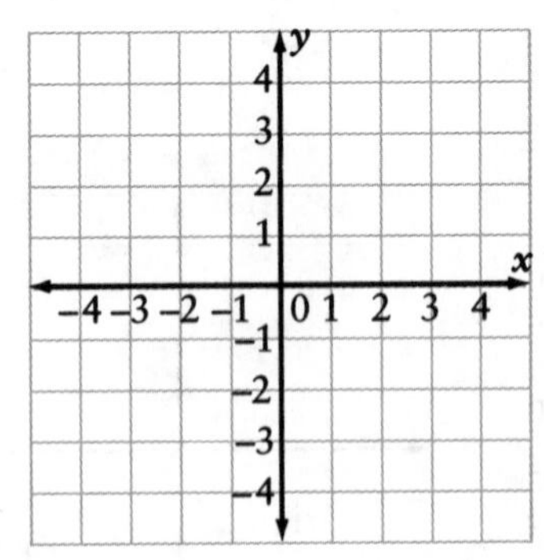

A (0, 2)
B (2, 4)
C (−3, −2)
D (4, 0)
E (−2, 2)
F (3, −1)

5.

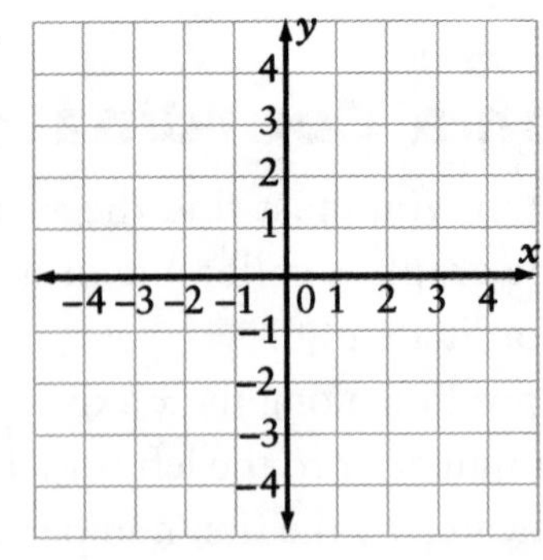

A (4, 1)
B (0, 0)
C (−3, 1)
D (−4, −4)
E (2, 2)
F (−3, −1)

6.

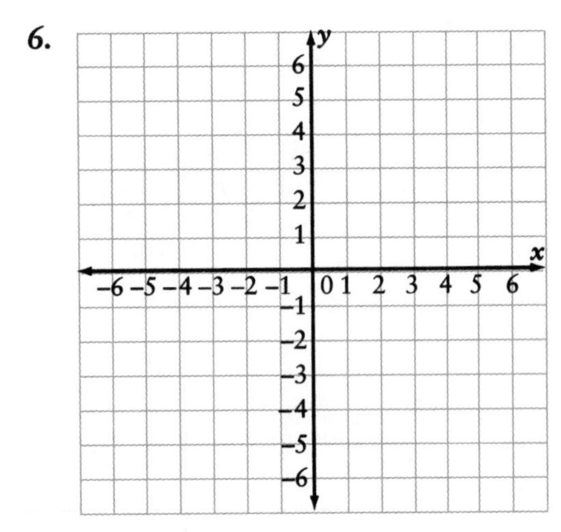

A (−6, 2)
B (0, 4)
C (3, −2)
D (4, 2)
E (−5, −3)
F (2, −6)

Graph the points. Then connect them with a line.

7.

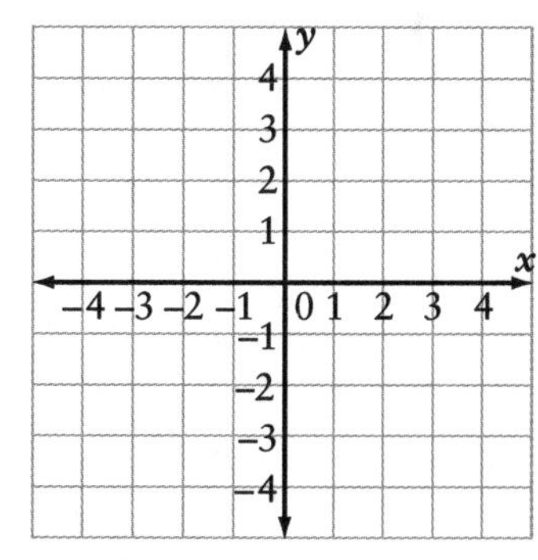

A (0, −3)
B (2, 0)

8.

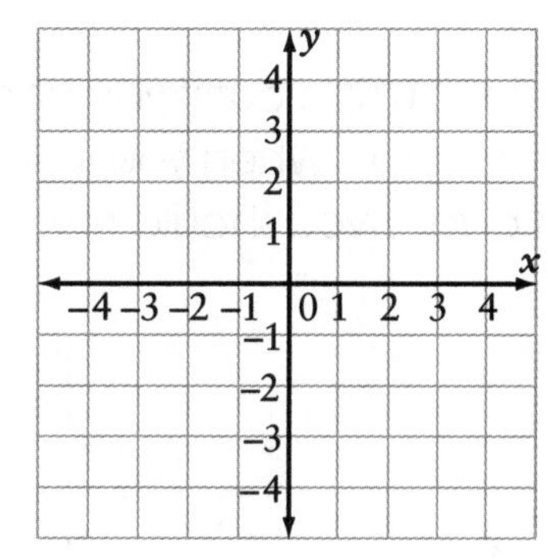

A (−1, 0)
B (1, 2)

9.

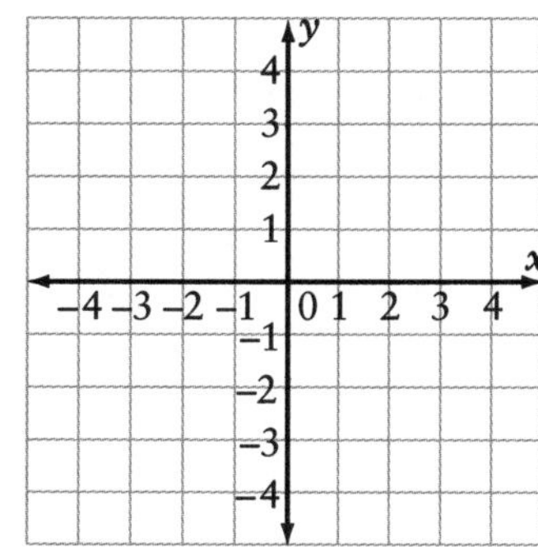

A (−2, 3)
B (0, 2)
C (2, 1)

To check your answers, turn to page 128.

COORDINATE PLANE REVIEW

The problems on this page will help you find out if you need to review the coordinate plane section of this book. When you finish, look at the chart on page 103 to see which pages you should review.

Use the graph to answer 1–6.

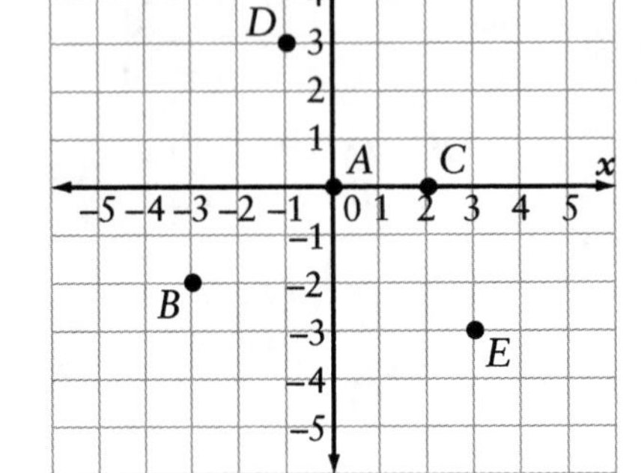

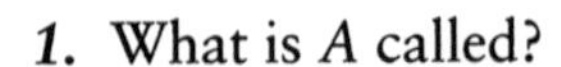

1. What is A called?

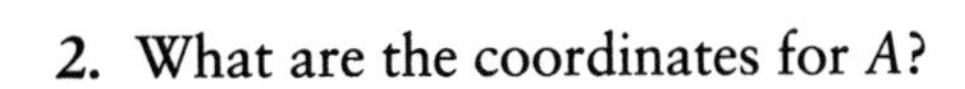

2. What are the coordinates for A?

3. What are the coordinates for B?

4. What are the coordinates for C?

5. What are the coordinates for D?

6. What are the coordinates for E?

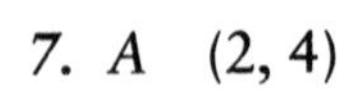

Plot the points on the graph.

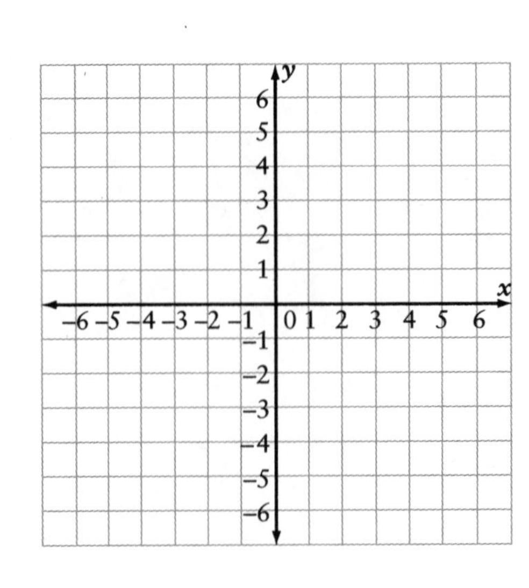

7. A $(2, 4)$

8. B $(-3, 0)$

9. C $(2, -2)$

10. D $(0, -1)$

11. E $(3, -6)$

12. F $(-1, -4)$

PROGRESS CHECK

Check your answers on page 129. Then return to the review pages for the problems you missed. Correct your answers before going on to the next unit.

If you missed problems	*Review pages*
1 to 6	96 to 98
7 to 12	99 to 101

FINAL REVIEW

These problems will tell you which sections of this book you need to study further. When you finish, check the chart to see which pages you need to review.

Name the kind of angle for items 1–5.

1. 

2. 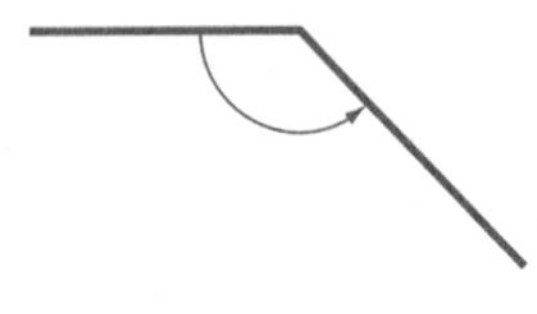

3. 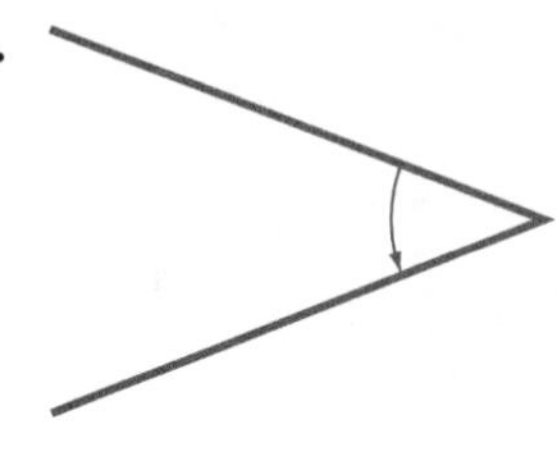

4. 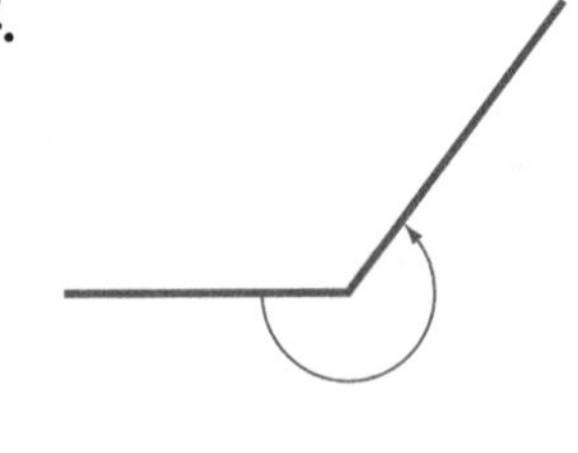

5. 

6. What is the complement of a 74° angle? ______________

7. What is the supplement of a 105° angle? ______________

Use the figure to answer 8–13.

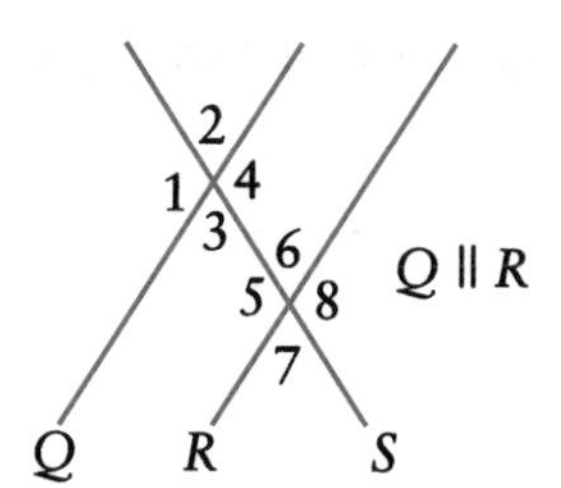

8. What type of line is S?

9. What is the vertical angle for $\angle 2$? __________________

10. What type of lines are Q and R? __________________

11. If $\angle 3$ is 60°, what is the measure of $\angle 4$? __________________

12. What is the corresponding angle for $\angle 6$? __________________

13. If $\angle 1$ is 112°, what is the measure of $\angle 2$? __________________

Name the kind of triangle for items 14–17.

14.

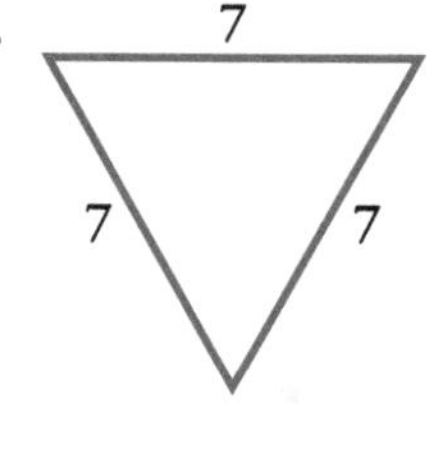

____

15.

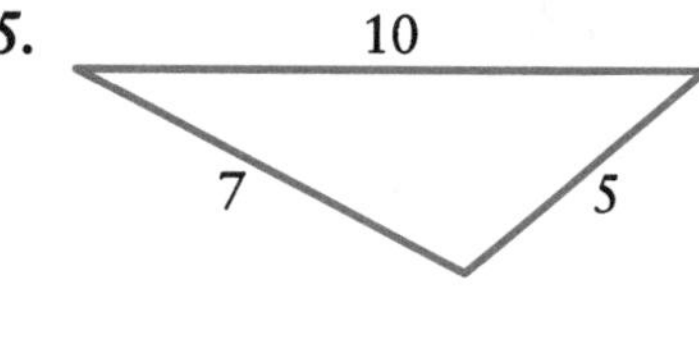

____

16.

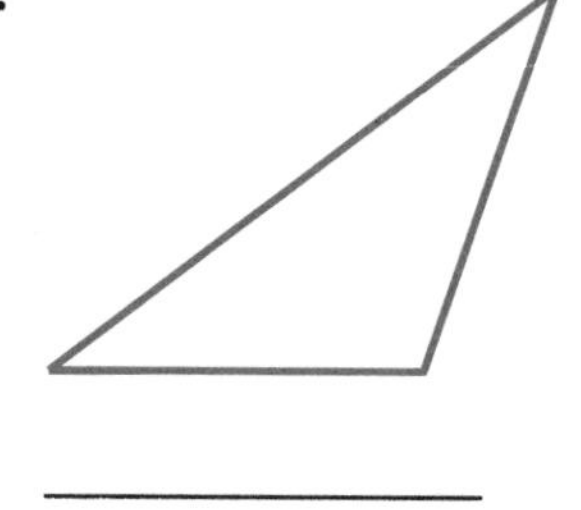

____

17.

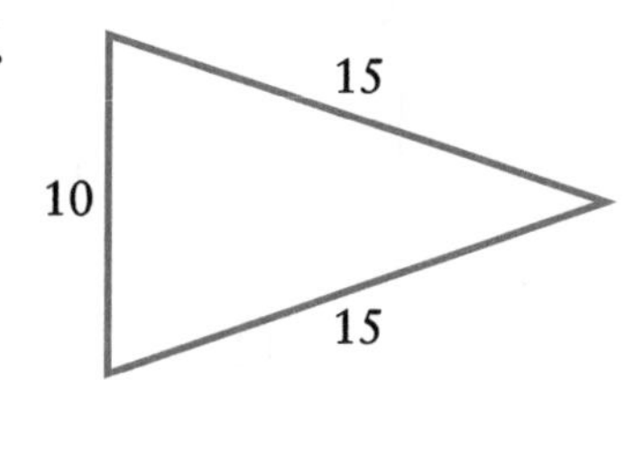

____

Find the area and perimeter of the triangle below.

18. $A =$ ________ **19.** $P =$ ________

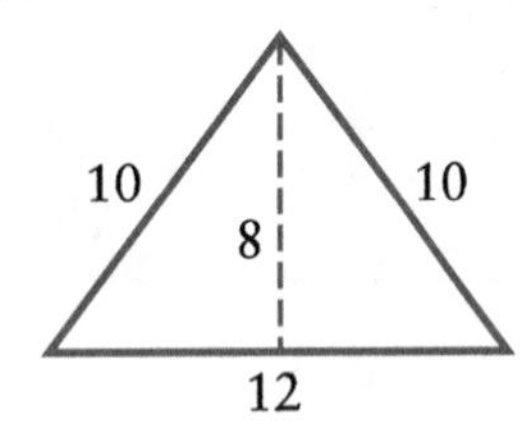

Find the length of the unknown side.

20.

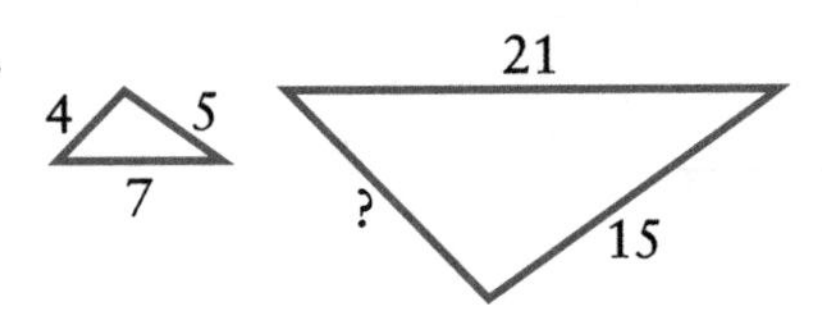

similar triangles

____

21.

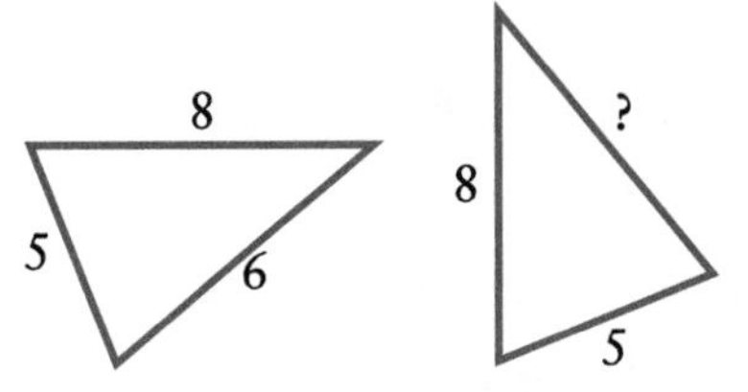

congruent triangles

____

Name the type of quadrilateral.

22.

8

3

____

23.

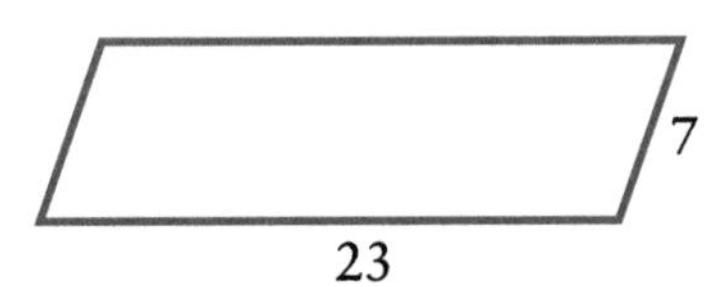

____

24.

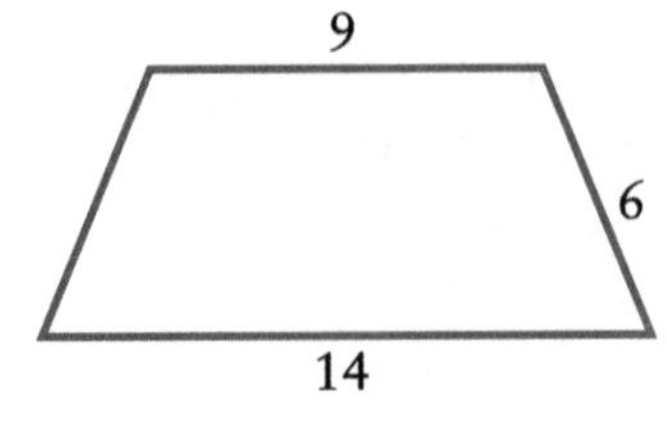

____

25.

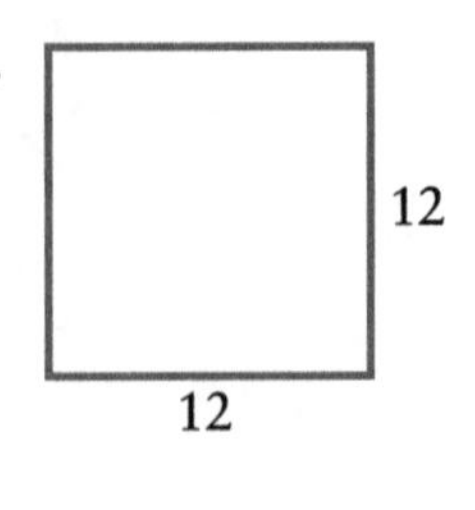

____

Find the area and perimeter (or circumference) of each.
Use 3.14 for π.

26.

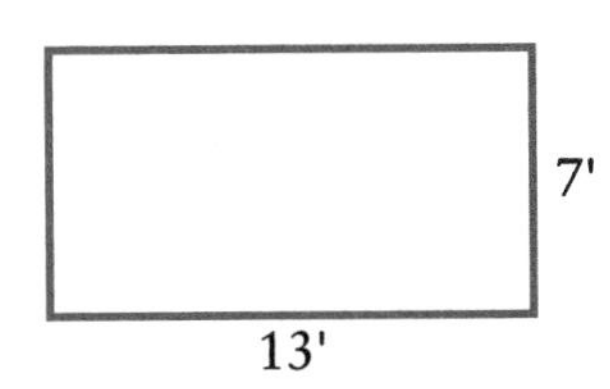

$A =$ ________

$P =$ ________

27.

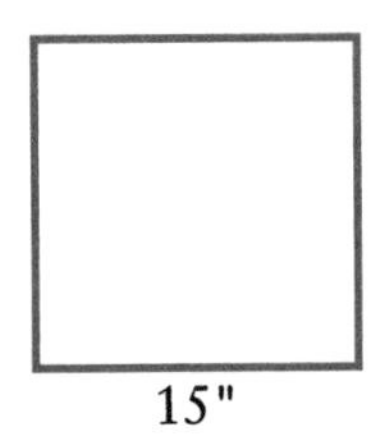

$A =$ ________

$P =$ ________

28.

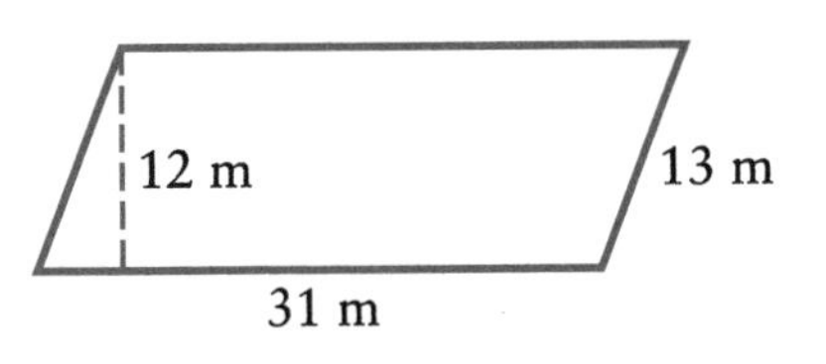

$A =$ ________

$P =$ ________

29.

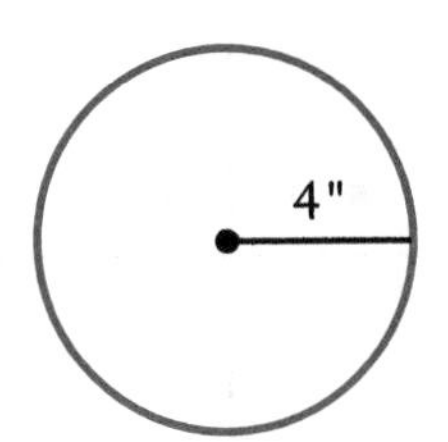

$A =$ ________

$C =$ ________

Name the type of solid. Find the volume of each solid.

30.

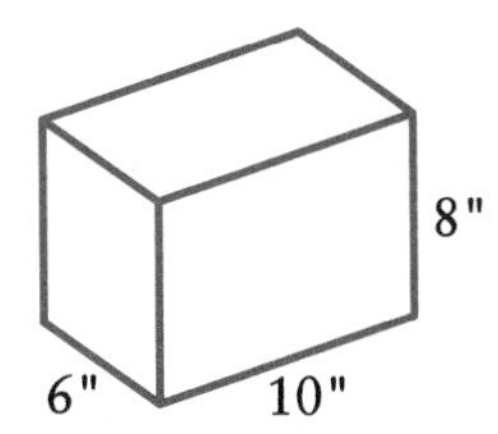

$l = 10"$
$w = 6"$
$h = 8"$

$V =$ ________

31.

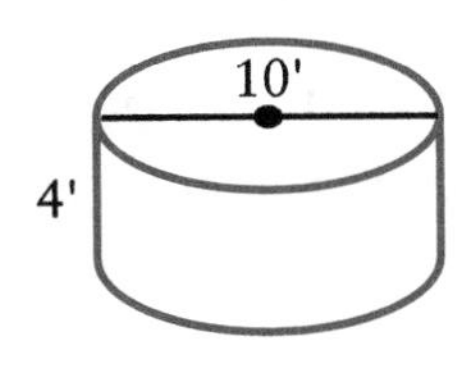

$d = 10'$
$h = 4'$

$V =$ ________

32.

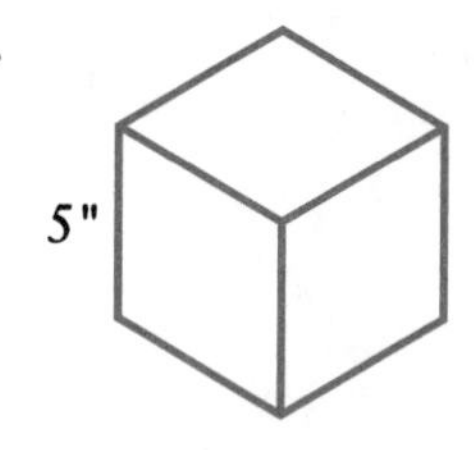

$s = 5"$

$V =$ ________

Write the coordinates for the points on the graph.

33. *A*

34. *B*

35. C

36. *D*

37. E

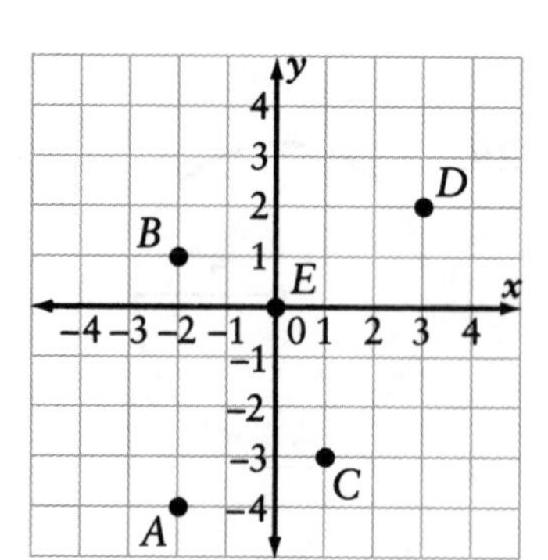

38. Jeremy wants to put a fence around his yard that is 60 feet by 84 feet. How much fencing will he need?

39. Jeremy plans to fertilize his 60-by-84-foot yard. One bag of fertilizer will cover 200 square yards. How many bags will Jeremy need to cover his yard? [**Hint:** Convert the dimensions of the yard from feet to yards.]

40. Berta wants to put an 8" base of concrete on her new patio. The patio is 10 ft by 15'. What is the volume of concrete she needs to order?

FINAL PROGRESS CHECK

Check your answers on page 129. Circle the problems you missed on the chart below. Review the pages that show how to work the problems you missed. Then try the problems again.

Problem Number	*Review Pages*	*Problem Number*	*Review Pages*	*Problem Number*	*Review Pages*
1	13–15	16	29–31	31	90–92
2	13–15	17	29–31	32	84–86
3	13–15	18	34–35	33	96–98
4	13–15	19	34–35	34	96–98
5	13–15	20	36–37	35	96–98
6	16	21	48–50	36	96–98
7	17–18	22	54–56	37	96–98
8	19–25	23	54–56	38	60–62
9	19–25	24	54–56	39	60–62
10	19–25	25	54–56	40	87–89
11	19–25	26	60–62		
12	19–25	27	57–59		
13	19–25	28	63–66		
14	29–31	29	70–74		
15	29–31	30	87–89		

Answers

Geometry Preview
pages 1–5

1. acute
2. obtuse
3. right
4. straight
5. reflex
6. 53°
7. 132°
8. $\angle 5$
9. parallel
10. transversal
11. $\angle 5$
12. 115°
13. 115°
14. isosceles
15. right
16. scalene
17. equilateral
18. 39 units
19. 75 sq units
20. 14
21. 11
22. trapezoid
23. square
24. rectangle
25. parallelogram
26. $A = 36$ sq units
    $P = 24$ units
27. $A = 84$ sq ft
    $P = 38$ ft
28. $A = 144$ sq in
    $P = 56$ in
29. $A = 78.5$ sq in
    $C = 31.4$ in
30. $A = 113.04$ sq cm
    $C = 37.68$ cm
31. cube
32. sphere
33. rectangular solid
34. cylinder
35. cone
36. 314 cubic inches
37. 720 cubic inches
38. 512 cubic cm
39. (0, 3)
40. (−2, 0)
41. (2, −2)
42. (0, 0)
43. (−3, −4)

Defining an Angle
Practice 1
pages 8–9

1. $\angle B$ or $\angle ABC$ or $\angle CBA$
2. $\angle 3$
3. $\angle XYZ$ or $\angle Y$ or $\angle ZYX$
4. $\angle$D
5. $\angle PQR$ or $\angle RQP$ or $\angle Q$
6. $\angle 7$

7–9 Answers will vary.

7.
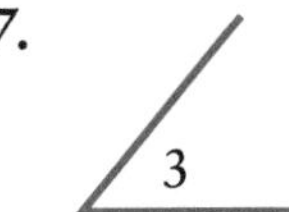

8.
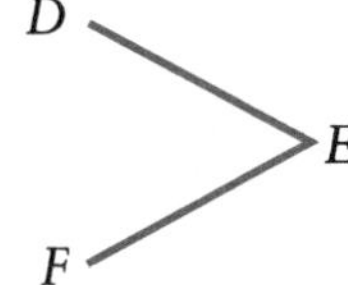

9.
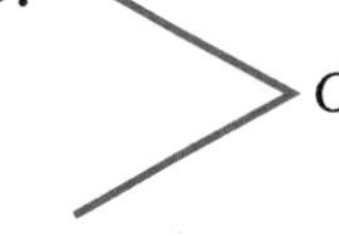

10. d
11. a
12. e
13. b
14. c
15. point of an angle
16. space defined by where two lines meet
17. Answers will vary. tile, door frames, window frames, boxes, etc.

Using a Protractor
Practice 2
pages 11–12

1. 60°
2. 120°
3. 40°
4. 80°
5. 35°
6. 110°
7. 90°
8. 60°
9. 50°
10. 40°
11. 120°
12. 90°
13. 65°
14. 45°

Kinds of Angles
Practice 3
pages 14–15

1. reflex
2. right
3. straight
4. acute
5. obtuse

6–10 Answers will vary.

6.

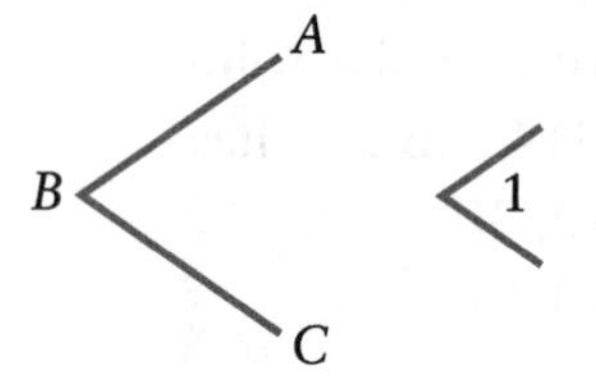

7.

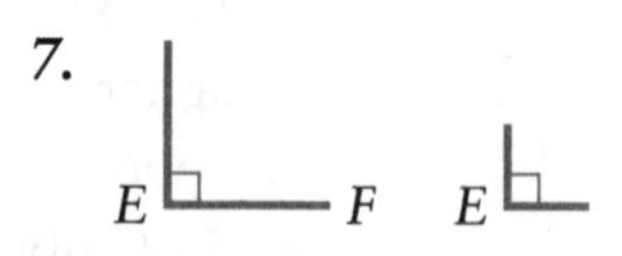

8.

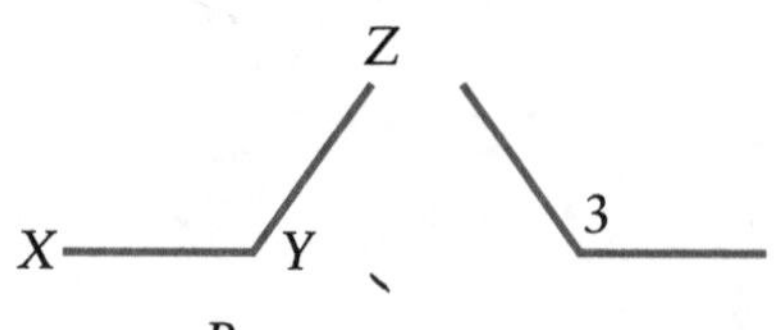

9. M N O Q

10.

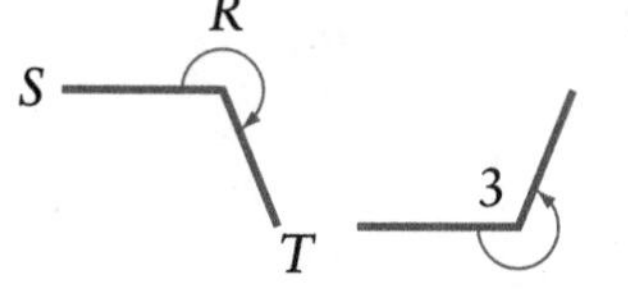

11. b
12. c
13. d
14. a
15. e
16. obtuse

Complementary Angles
Practice 4
page 16

1. 30°
2. 87°
3. 10°
4. 20°
5. 64°
6. 1°
7. Angle 46°
Complement 44°
8. Angle 38°
Complement 52°
9. Angle 85°
Complement 5°
10. Angle 18°
Complement 72°
11. 25°

Supplementary Angles
Practice 5
pages 17–18

1. 180− 40 = 140°
2. 180 − 72 = 108°
3. 180 − 135 = 45°
4. 180 − 120 = 60°
5. 180 − 97 = 83°
6. 180 − 5 = 175°
7. Angle 130°
   Supplement 50°
8. ∠90°
   Supplement 90°
9. Angle 60°
   Supplement 120°
10. c
11. a
12. b
13. b
14. 80°
15. 130°

Parallel lines
Practice 6
pages 19–20

1. no
2. yes
3. yes
4. no
5. no

6–7 Answers will vary.

6. door frames, window frames, counter tops, shelves, tiles, carpet runners, etc.
7. railroad tracks, electric wires, streets, telephone poles, fence posts, football goal posts, etc.
8. 
9. 
10.
11. yes
12. no
13. yes
14. no

Perpendicular Lines
Practice 7
pages 21–22

1. no
2. yes
3. yes
4. no
5. no
6. yes

7–8 Answers will vary.

7. corners of a box, corners of a tile, corners of window and door frames
8. skyline, poles intersecting the ground, corners of billboards

9.

10. 

11. 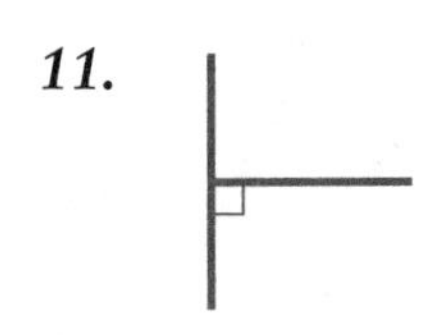

12. yes
13. no
14. no
15. yes
16. no
17. yes

Transversals and Vertical Angles
Practice 8
pages 24–25

1. $\angle g$
2. $\angle a$
3. $\angle f$
4. parallel
5. $\angle f$
6. $\angle a$
7. $\angle b$
8. $\angle g$, $\angle c$, $\angle b$
9. $\angle d$, $\angle e$, $\angle h$
10. $\angle a = 120°$, $\angle b = 60°$, $\angle c = 60°$, $\angle d = 120°$, $\angle f = 60°$, $\angle g = 60°$, $\angle h = 120°$
11. $\angle 1 = 130°$, $\angle 2 = 50°$, $\angle 3 = 50°$, $\angle 5 = 130°$, $\angle 6 = 50°$, $\angle 7 = 50°$, $\angle 8 = 130°$
12. $\angle 1 = 85°$, $\angle 2 = 95°$, $\angle 3 = 95°$, $\angle 4 = 85°$, $\angle 6 = 95°$, $\angle 7 = 95°$, $\angle 8 = 85°$
13. $\angle 1$, $\angle 4$, $\angle 5$
14. $\angle 2$, $\angle 3$, $\angle 6$
15. $\angle 6$
16. $\angle 4$
17. 120°
18. $\angle 2 = 130°$, $\angle 3 = 130°$, $\angle 4 = 50°$, $\angle 5 = 50°$, $\angle 6 = 130°$, $\angle 7 = 130°$, $\angle 8 = 50°$

Angles & Lines Review
pages 26–28

1. $\angle DEF$ or $\angle FED$ or $\angle E$
2. $\angle 1$
3. $\angle$a
4. acute
5. right
6. obtuse
7. reflex
8. straight
9. acute
10. $90 - 25 = 65°$
11. $90 - 84 = 6°$
12. $90 - 51 = 39°$
13. $180 - 30 = 150°$
14. $180 - 62 = 118°$
15. $180 - 115 = 65°$
16. $90 - 36 = 54°$
    $180 - 36 = 144°$
17. $90 - 74 = 16°$
    $180 - 74 = 106°$
18. $90 - 58 = 32°$
    $180 - 58 = 122°$
19. c
20. g
21. k
22. a
23. i
24. l

25. h
26. j
27. e
28. f
29. b
30. d
31. parallel
32. perpendicular
33. transversal
34. $\angle 5$
35. $\angle 7$
36. 90°
37. $\angle 5$, $\angle 9$, $\angle 12$
38. 110°

Kinds of Triangles
Practice 9
pages 30–32

1. b
2. c
3. a
4. d
5. equilateral
6. isosceles
7. scalene
8. right
9. equilateral
10. scalene
11. isosceles
12. right
13. right
14. 1", 1", 1" equilateral
15. 1", $1\frac{1}{4}$", $1\frac{1}{4}$" isosceles
16. 1", $\frac{3}{4}$", $1\frac{1}{4}$" right
17. $1\frac{13}{16}$", $1\frac{1}{8}$", $\frac{7}{8}$" scalene
18. isosceles
19. scalene

The Angles in a Triangle
Practice 10
page 33

1. $180 - 52 - 74 = 54°$
2. $180 - 90 - 55 = 35°$
3. $180 - 37 - 48 = 95°$
4. $180 - 106 - 28 = 46°$
5. $180 - 70 - 70 = 40°$
6. $180 - 59 - 32 = 89°$
7. 42°
8. $\angle P = 70°$
   $\angle Q = 70°$
9. $180 - 62 - 62 = 56°$

The Perimeter and Area of Triangles
Practice 11
pages 35–36

A.

1. $A = \frac{1}{2} \cdot 10 \cdot 15 = 75$ sq cm
   $P = 16 + 17 + 10 = 43$ cm
2. $A = \frac{1}{2} \cdot 6 \cdot 8 = 24$ sq m
   $P = 10 + 8 + 6 = 24$ m
3. $A = \frac{1}{2} \cdot 17 \cdot 10 = 85$ sq in
   $P = 11 + 17 + 14 = 42$ in
4. $A = \frac{1}{2} \cdot 12 \cdot 9 = 54$ sq mm
   $P = 12 + 12 + 12 = 36$ mm

B.

5. $A = \frac{1}{2} \cdot 1 \cdot 2 = 1$ sq ft or $A = \frac{1}{2} \cdot 12 \cdot 24 = 144$ sq in
   $P = 14 + 16 + 24 = 54$ in
6. $A = \frac{1}{2} \cdot 24 \cdot 30 = 360$ sq in or $A = \frac{1}{2} \cdot 2 \cdot 2\frac{1}{2} = 2\frac{1}{2}$ sq ft
   $P = 3 + 3 + 2 = 8$ ft or $P = 36" + 36" + 24" = 96"$
7. $A = \frac{1}{2} \cdot 10 \cdot 18 = 90$ sq ft
   $P = 12 + 14 + 18 = 44$ ft
8. $A = \frac{1}{2} \cdot 36 \cdot 18 = 324$ sq in
   $P = 24 + 27 + 36 = 87$ in

Similar Triangles
Practice 12
pages 37–38

1. $\frac{12}{8} = \frac{9}{?}$ $\quad ? = 6$
2. $\frac{15}{20} = \frac{12}{?}$ $\quad ? = 16$
3. $\frac{7}{14} = \frac{12}{?}$ $\quad ? = 24$
4. $\frac{3}{9} = \frac{16}{?}$ $\quad ? = 48$
5. $\frac{9}{18} = \frac{6}{?}$ $\quad \overline{RS} = 12$
   $\frac{9}{18} = \frac{12}{?}$ $\quad \overline{RT} = 24$
6. $\frac{5}{10} = \frac{?}{13}$ $\quad \overline{AD} = 6\frac{1}{2}$
   $\frac{5}{10} = \frac{?}{18}$ $\quad \overline{DB} = 9$

7. $\frac{3}{60} = \frac{?}{100}$ $\overline{AC} = 5$

$\frac{3}{60} = \frac{?}{80}$ $\overline{AB} = 4$

8. $\frac{24}{120} = \frac{?}{85}$? = 17 inches

$\frac{24}{120} = \frac{?}{100}$? = 20 in

Equilateral Triangles
Practice 13
pages 39–40

1. 60
2. same
3. 180
4. $A = \frac{1}{2} \cdot 5 \cdot 6 = 15$ sq yd

   $P = 6 + 6 + 6 = 18$ yd
5. $A = \frac{1}{2} \cdot 12 \cdot 16 = 96$ sq in

   $P = 16 + 16 + 16 = 48$ in
6. $A = \frac{1}{2} \cdot 25 \cdot 31 = 387.5$ sq m

   $P = 31 + 31 + 31 = 93$ m
7. $A = \frac{1}{2} \cdot \frac{1}{2} \cdot \frac{1}{3} = \frac{1}{12}$ sq ft or $\frac{1}{2} \cdot 4 \cdot 6 = 12$ sq in

   $P = \frac{1}{2} + \frac{1}{2} + \frac{1}{2} = 1\frac{1}{2}$ ft
8. $30 = 3s$

   $10 = s$
9. $63 = 3s$

   $21 = s$
10. $12 = 3s$

    $4 = s$
11. $21.9 = 3s$

    $7.3 = s$
12. $120 = \frac{1}{2} \cdot 12 \cdot b$

    $20 = b$
13. $75 = \frac{1}{2} \cdot 10 \cdot b$

    $15 = b$
14. $48 = \frac{1}{2} \cdot 8 \cdot b$

    $12 = b$
15. $75 = 3s$

    $25 = s$
16. $80 = \frac{1}{2} \cdot 10 \cdot b$

    $16 = b$

Isosceles Triangles
Practice 14
pages 41–43

1. Yes
2. Yes
3. No
4. Yes
5. No
6. No
7. $180 - 50 = 130$ $130 \div 2 = 65°$
8. $180 - 70 = 110$ $110 \div 2 = 55°$
9. $180 - 20 = 160$ $160 \div 2 = 80°$
10. $180 - 36 = 144$ $144 \div 2 = 72°$
11. $180 - 62 - 62 = 56°$
12. $180 - 51 - 51 = 78°$
13. $180 - 73 - 73 = 34°$
14. $180 - 26 - 26 = 128°$
15. $180 - 68 = 112$ $112 \div 2 = 56°$

Right Triangles
Practice 15
pages 44–45

1. 90°
2. hypotenuse
3. 180°
4. Yes
5. No
6. Yes
7. No
8. $90 - 40 = 50°$
9. $90 - 32 = 58°$
10. $90 - 45 = 45°$
11. $90 - 20 = 70°$
12. $90 - 40 = 50°$

Pythagorean Theorem
Practice 16
pages 47–48

1. hypotenuse
2. legs
3. Pythagorean
4. right
5. acute
6. $6^2 + 8^2 = c^2$
   $c = 10$
7. $17^2 = 8^2 + a^2$
   $15 = a$
8. $c^2 = 12^2 + 5^2$
   $c = 13$
9. $30^2 + 40^2 = c^2$
   $50 = c$

10. $c^2 = 6^2 + 8^2$
$c = 10$
$P = 6 + 8 + 10 = 24'$

11. $25^2 = 15^2 + b^2$
$20 = b$

12. $3^2 + 4^2 = c^2$
$5 = c$

13. $13^2 = 5^2 + b^2$
$12 = b$

Congruent Triangles
Practice 17
pages 49–50

1. Yes
2. No
3. Yes
4. No
5. No
6. Yes
7. 11'
8. 6"
9. $\angle A$ and $\angle Y$, $\angle B$ and $\angle Z$, $\angle C$ and $\angle X$
10. 8
11. 6
12. 11
13. 29°
14. $180 - 29 - 37 = 114°$
15. 114°

Triangles Review
pages 51–53

1. 180
2. 3
3. $\frac{1}{2}bh = A$
4. $P = a + b + c$
5. Pythagorean
6. d
7. e
8. f
9. c
10. b
11. a
12. right
13. equilateral
14. isosceles
15. scalene
16. right
17. $180 - 90 - 30 = 60°$
18. $180 - 114 - 35 = 31°$
19. $A = \frac{1}{2} \cdot 24 \cdot 20 = 240$ sq in
$P = 26 + 26 + 20 = 72$ inches
20. $A = \frac{1}{2} \cdot 12 \cdot 24 = 144$ sq cm
$P = 16 + 18 + 24 = 58$ cm

21. $A = \frac{1}{2} \cdot 12 \cdot 8 = 48$ sq units
$P = 10 + 10 + 12 = 32$ units
22. $A = \frac{1}{2} \cdot 5 \cdot 12 = 30$ sq units
$P = 12 + 5 + 13 = 30$ units
23. $10^2 = 6^2 + a^2$
$8 = a$
24. $15^2 + 8^2 = c^2$
$17 = c$
25. 4" and 5"
26. $5^2 + 12^2 = c^2$
$13 = c$
27. $60 = 3s$
$20 = s$

Unit 3. Kinds of Quadrilaterals
Practice 18
pages 55–57

1. parallelogram
2. square
3. rectangle
4. trapezoid
5. rectangle
6. square
7. true
8. false
9. true
10. false
11. true
12. false
13. $\angle A = 90°$
14. $\angle A = 98°$ $\angle C = 98°$
15. $\angle B = 90°$
16. $\angle A = 130°$

Perimeter and Area of Squares
Practice 19
pages 58–59

1. $A = 4^2 = 16$ sq ft
$P = 4 \cdot 4 = 16$ ft
2. $A = (2.5)^2 = 6.25$ sq m
$P = 4 \cdot 2.5 = 10$ m
3. $A = 1^2 = 1$ sq yd
$P = 4 \cdot 1 = 4$ yd
4. $A = (6.5)^2 = 42.25$ sq ft
$P = 4(6.5) = 26$ ft
5. $A = 80^2 = 6400$ sq blocks
$P = 4 \cdot 80 = 320$ blocks

6. $A = 16^2 = 256$ sq in
$P = 16 \cdot 4 = 64$ in
7. $A = \left(1\frac{3}{4}\right)^2 = 3\frac{1}{16}$ sq ft
$P = 4\left(1\frac{3}{4}\right) = 7$ ft
8. $A = 25^2 = 625$ sq ft
$P = 25 \cdot 4 = 100$ ft
9. $P = 30 \cdot 4 = 120"$ for 1 window
$120 \times 4 = 480"$ for 4 windows
10. $P = 8 \cdot 4 = 32$ inches
$32 \times 5 = 160$ inches for 5 frames
11. $A = 18^2 = 324$ sq in
12. $A = (250)^2 = 62{,}500$ sq yd
13. $P = 14 \cdot 4 = 56$ inches
$A = 14^2 = 196$ sq in

Perimeter and Area of Rectangles
Practice 20
pages 61–63

A.

1. $A = 12 \cdot 7 = 84$ sq ft
$P = 2 \cdot 12 + 2 \cdot 7 = 38$ ft
2. $A = 12 \cdot 8\frac{1}{2} = 102$ sq units
$P = 2 \cdot 12 + 2 \cdot 8\frac{1}{2} = 41$ units
3. $A = 6.4(3.5) = 22.4$ sq m
$P = 2(6.4) + 2(3.5) = 19.8$ m
4. $A = 3\frac{1}{2}\left(1\frac{1}{4}\right) = 4\frac{3}{8}$ sq yd
$P = 2\left(3\frac{1}{2}\right) + 2\left(1\frac{1}{4}\right) = 9\frac{1}{2}$ yd
5. $A = 8 \cdot 15 = 120$ sq m
$P = 2 \cdot 8 + 2 \cdot 15 = 46$ m
6. $A = 2 \cdot 12 = 24$ sq cm
$P = 2 \cdot 12 + 2 \cdot 2 = 28$ cm

B.

7. $A = 24 \cdot 35 = 840$ sq in
$P = 2 \cdot 24 + 2 \cdot 35 = 118$ in
8. $A = 2 \cdot 9 = 18$ sq ft
$P = 2 \cdot 2 + 2 \cdot 9 = 22$ ft
9. $A = 4 \cdot 7\frac{1}{2} = 30$ sq ft
$P = 2 \cdot 4 + 2 \cdot 7\frac{1}{2} = 23$ ft

10. $A = 18 \cdot 72 = 1{,}296$ sq in
    $P = 2 \cdot 18 + 2 \cdot 72 = 180$ in
11. $A = 16 \cdot 21 = 336$ sq ft
    $P = 2 \cdot 16 + 2 \cdot 21 = 74$ ft
12. $A = 40 \cdot 240 = 9{,}600$ sq in
    $P = 2 \cdot 40 + 2 \cdot 240 = 560$ in
13. $A = 14 \cdot 12 = 168$ sq ft
14. $P = 2 \cdot 14 + 2 \cdot 12 = 52$ ft

Perimeter and Area of Parallelograms
Practice 21
pages 64–66

1. Yes
2. No
3. No
4. No
5. $A = 9 \cdot 16 = 144$ sq ft
   $P = 2 \cdot 16 + 2 \cdot 11 = 54$ ft
6. $A = 5 \cdot 7 = 35$ sq in
   $P = 2 \cdot 7 + 2 \cdot 6 = 26$ in
7. $A = 1.8 \cdot 2.7 = 4.86$ sq m
   $P = 2 \cdot 2.7 + 2 \cdot 2.1 = 9.6$ m
8. $A = 3\frac{1}{2} \cdot 1\frac{3}{4} = 6\frac{1}{8}$ sq yd
   $P = 2 \cdot 3\frac{1}{2} + 2 \cdot 2\frac{1}{4} = 11\frac{1}{2}$ yd
9. $A = 9 \cdot 15 = 135$ sq in
   $P = 2 \cdot 12 + 2 \cdot 15 = 54$ in
10. $A = 4 \cdot 10 = 40$ sq ft
    $P = 2 \cdot 10 + 2 \cdot 7\frac{1}{2} = 35$ ft
11. $A = 3 \cdot 6 = 18$ sq ft
    $P = 2 \cdot 6 + 2 \cdot 4\frac{1}{2} = 21$ ft
12. $A = 5 \cdot 12 = 60$ sq ft
    $P = 2 \cdot 8 + 2 \cdot 12 = 40$ ft
13. $A = 14 \cdot 24 = 336$ sq ft
    $336 \div 75 = 4.48$ qts $= 1$ gal 1 qt
14. $P = 2 \cdot 6 + 2 \cdot 8 = 28$ yds

Quadrilaterals Review
pages 67–69

1. square
2. trapezoid
3. rectangle
4. parallelogram
5. rectangle
6. trapezoid
7. $A = 9 \times 14 = 126$ sq ft
   $P = 2 \cdot 9 + 2 \cdot 14 = 46$ ft
8. $A = 16 \times 12 = 192$ sq yd
   $P = 2 \cdot 16 + 2 \cdot 12 = 56$ yds
9. $A = 16 \times 16 = 256$ sq in
   $P = 16 \times 4 = 64$ in
10. $A = 8 \times 14 = 112$ sq units
    $P = 2 \cdot 10 + 2 \cdot 14 = 48$ units
11. 3 yd = 9'
    $A = 2 \times 9 = 18$ sq ft
    $P = 2 \cdot 2 + 2 \cdot 9 = 22$ ft
12. 4 yd = 12'
    $A = 12 \times 22 = 264$ sq ft
    $P = 2 \cdot 22 + 2 \cdot 13 = 70$ ft
13. $A = 60 \times 60 = 3{,}600$ sq cm
    $P = 60 \times 4 = 240$ cm
14. 18" $= \frac{1}{2}$ yd
    $A = \frac{1}{2} \cdot 2 = 1$ sq yd
    $P = 2 \cdot 2 + 2 \cdot \frac{1}{2} = 5$ yd
15. 24" = 2'
    $P = 2 \cdot 2 + 2 \cdot 3\frac{1}{2} = 4 + 7 = 11'$
    for 3 windows $3 \times 11 = 33'$
16. 6' = 72"
    $A = 20 \times 72 = 1440$ sq in
    1500 is enough to do the job.

Finding Circumference
Practice 22
pages 71–73

1. circumference
2. radius
3. diameter
4. 3.14
5. $C = 2\pi r$ or $C = \pi d$
6. $2 \times 2 = 4"$
7. $4' \times 2 = 8'$
8. $6 \times 2 = 12$ m
9. $10 \times 2 = 20'$
10. $1 \times 2 = 2$ yd
11. $4.2 \times 2 = 8.4$ cm
12. $12 \div 2 = 6$ yd
13. $6.4 \div 2 = 3.2$ m
14. $4.5 \div 2 = 2.25'$
15. $21 \div 2 = 10.5"$
16. $16 \div 2 = 8$ mm

17. $4.6 \div 2 = 2.3$ m

18. $2 \cdot 4 \cdot 3.14 = 25.12'$

19. $2 \cdot 2 \cdot 3.14 = 12.56"$

20. $2 \cdot 7 \cdot 3.14 = 43.96$ yd

21. $2 \cdot 5.5 \cdot 3.14 = 34.54"$

22. $9.6 \cdot 3.14 = 30.144$ m

23. $2 \cdot 6.8 \cdot 3.14 = 42.704$ cm

24. $r = 10'$
$C = 2 \cdot 10 \cdot 3.14 = 62.8'$

25. $48 \cdot 3.14 = 150.72"$
$5' = 60"$
$150.72 \div 60 = 2+$
She buys 3 packages.

26. $120 \cdot 3.14 = 376.8$ yd
$376.8 \times 8 = 3014.4$ yd

Finding Area
Practice 23
pages 74–75

1. square
3. radius
5. $10 \div 2 = 5'$
7. $9.5 \div 2 = 4.75$ yd
9. $7^2 \cdot 3.14 = 153.86$ sq ft
11. $8^2 \cdot 3.14 = 200.96$ sq m
13. $9.5^2 \cdot 3.14 = 283.385$ sq in
14. $12^2 \cdot 3.14 = 452.16$ sq in
15. $4^2 \cdot 3.14 = 50.24$ sq ft
16. $36 \div 2 = 18$ inches
$18^2 \cdot 3.14 = 1{,}017.36$ sq inches

Circles Review
pages 76–78

1. circumference
2. diameter
3. radius
4. $C = 2\pi r$ or $c = \pi d$
5. $A = \pi r^2$
6. $10 \div 2 = 5'$
7. $16.4 \div 2 = 8.2$ m
8. $12.6 \div 2 = 6.3'$
9. $6 \cdot 2 = 12"$
10. $5 \cdot 2 = 10$ cm
11. $2.4 \cdot 2 = 4.8$ m
12. $A = 3.14 \cdot 6^2 = 113.04$ sq ft
13. $A = 3.14 \cdot 4^2 = 50.24$ sq in
14. $A = 3.14 \cdot 2.5^2 = 19.625$ sq ft
15. $A = 3.14 \cdot 5^2 = 78.5$ sq yd
16. $A = 3.14 \cdot .9^2 = 2.5434$ sq m
17. $A = 3.14 \cdot 10^2 = 314$ sq ft
18. $C = 3.14 \cdot 2 \cdot 5 = 31.4$ ft
19. $C = 3.14 \cdot 2 \cdot 2.4 = 15.072$ m
20. $C = 3.14 \cdot 2 \cdot 10 = 62.8"$
21. $C = 3.14 \cdot 2 \cdot 6 = 37.68$ yd
22. $C = 3.14 \cdot 14 = 43.96$ ft
23. $C = 3.14 \cdot 80 = 251.2$ cm
24. $3.14 \cdot 20 = 62.8$ ft
25. $A = 3.14 \cdot 10^2 = 314$ sq ft
$314 \div 100 = 3.14$
4 quarts are needed.

Definitions of Solids
Practice 24
pages 80–81

1. cylinder
2. cube
3. rectangular solid
4. sphere
5. rectangular solid
6. sphere
7. rectangular solid or cube
8. cone
9. cylinder
10. rectangular solid
11. rectangular solid
12. cylinder

Definition of Volume
Practice 25
pages 82–83

1. volume
2. three
3. cubic

4–5 Answers will vary.

4. packing boxes, water in washer, amount of gas used, etc.
5. water to fill a swimming pool, dirt removed from a hillside, gas to fill tanker
6. yes
7. no
8. yes
9. no
10. yes
11. no
12. yes
13. yes
14. yes

Volume of a Cube
Practice 26
pages 85–86

1. $1.5 \times 1.5 \times 1.5 = 1.5^3 = 3.375$ cu ft
2. $4^3 = 64$ cu cm
3. $6^3 = 216$ cu ft
4. $15^3 = 3{,}375$ cu cm
5. $\left(\frac{3}{4}\right)^3 = \frac{27}{64}$ cu in
6. $2.5^3 = 15.625$ cu yd
7. $10^3 = 1{,}000$ cu ft
8. $2^3 = 8$ cu in
9. $5^3 = 125$ cu meters
10. $3^3 = 27$ cu in
11. $4.5^3 = 91.125$ cu ft

Volume of a Rectangular Solid
Practice 27
pages 87–90

A.

1. $V = 6 \times 3 \times 5 = 90$ cu in
2. $V = 4 \times 2 \times 8 = 64$ cu ft
3. $V = 14 \times 7 \times 10 = 980$ cu ft
4. $V = 6 \times 3 \times 12 = 216$ cu yd
5. $V = 3 \times 2.4 \times 4.1 = 29.52$ cu m
6. $V = 8 \times 6 \times 4.5 = 216$ cu cm

B.

7. $1\frac{1}{2}' = 18"$
   $V = 10 \times 6 \times 18 = 1{,}080$ cu in
8. $27" = \frac{3}{4}$ yd
   $V = 1\frac{1}{2} \times \frac{3}{4} \times 2 = \frac{9}{4} = 2\frac{1}{4}$ cu yd
9. $30" = 2\frac{1}{2}'$
   $V = 2\frac{1}{2} \times 2 \times 4 = 20$ cu ft
10. 1 yd = 36"
    $V = 18 \times 8 \times 36 = 5{,}184$ cu in
11. 2 yd = 6'
    $V = 4 \times 2 \times 6 = 48$ cu ft
12. $30" = 2\frac{1}{2}$ ft
    $V = 2\frac{1}{2} \times 2 \times 3\frac{1}{2} = 17\frac{1}{2}$ cu ft
13. $V = 12 \times 4 \times 8 = 384$ cu in
14. 12' = 4 yd
    $V = 3 \times 4 \times 5 = 60$ cu yd
15. $V = 3 \times 8 \times 11 = 264$ cu in

Volume of a Cylinder
Practice 28
pages 91–92

1. $V = 3.14 \cdot 2^2 \cdot 5 = 62.8$ cu in
2. $V = 3.14 \cdot 1^2 \cdot 6 = 18.84$ cu yd
3. $V = 3.14 \cdot 10^2 \cdot 3 = 942$ cu ft
4. $V = 3.14 \cdot 4^2 \cdot 2 = 100.48$ cu m
5. $V = 3.14 \cdot 6^2 \cdot 5.5 = 621.72$ cu ft
6. $V = 3.14 \cdot 5^2 \cdot 7 = 549.5$ cu in
7. 36" = 3'
   $V = 3.14 \cdot 2^2 \cdot 3 = 37.68$ cu ft

8. $2' = 24"$
$V = 3.14 \cdot 18^2 \cdot 24 = 24{,}416.64$ cu in

9. $2\frac{1}{2}' = 30"$
$V = 3.14 \cdot 10^2 \cdot 30 = 9{,}420$ cu in

10. $36" = 3'$
$V = 3.14 \cdot 3^2 \cdot 3 = 84.78$ cu ft

11. $V = 3.14 \cdot 10^2 \cdot 18 = 5{,}652$ cu ft

12. $V = 3.14 \cdot 2^2 \cdot 4 = 50.24$ cu ft

13. $50.24 \times 5 = 251.2$ lbs

Solids Review
pages 93–95

1. rectangular solid
2. sphere
3. cylinder
4. cube
5. c
6. b
7. a
8. $V = 6 \times 4 \times 5 = 120$ cu ft
9. $V = 3.14 \times 4^2 \times 6 = 301.44$ cu ft
10. $V = 8^3 = 512$ cu in
11. $V = 7 \times 3.5 \times 10 = 245$ cu in
12. $V = 6^2 = 216$ cu m
13. $24" = 2'$
$V = 3.14 \times 2^2 \times 5 = 62.8$ cu ft
14. $V = 3.14 \times 5^2 \times 2 = 157$ cu ft
15. $V = 12 \times 6 \times 7 = 504$ cu in
16. $18" = 1\frac{1}{2}'$
$V = 1 \times 1\frac{1}{2} \times 3 = 4\frac{1}{2}$ cu ft
17. $V = 2\frac{1}{4} \times 3 \times 4 = 27$ cu ft
18. $V = 3.14 \times 1.5^2 \times 2 = 14.13$ cu m
19. $9" = \frac{1}{4}$ yd
$V = 12 \times 5 \times \frac{1}{4} = 15$ cu yd

Unit 6. Ordered Pairs
Practice 29
page 98

1. $x = -2, y = 3$
2. $x = 0, y = 4$
3. $x = -3, y = -2$
4. $x = 2, y = -4$
5. $x = 12, y = 8$
6. $x = 3, y = -2$

7. x axis

8. y axis

9. -3

10. 0

11. $(-3, 0)$

12. $(-1, 3)$

13. $(1, 1)$

14. $(3, 2)$

15. $(4, -1)$

16. $(1, -3)$

17. $(-2, -2)$

18. $(-5, -3)$

19. $(-5, 4)$

20. $(3, -5)$

Plotting Pairs on a Coordinate Graph
Practice 30
pages 99–101

1. left or right
2. up or down
3. (0, 0)
4.

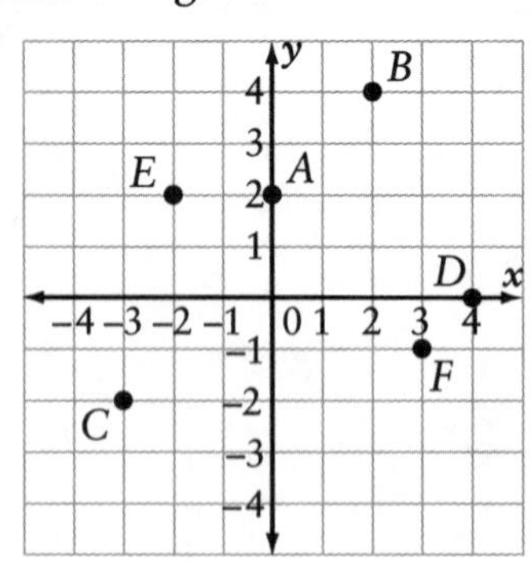

5.

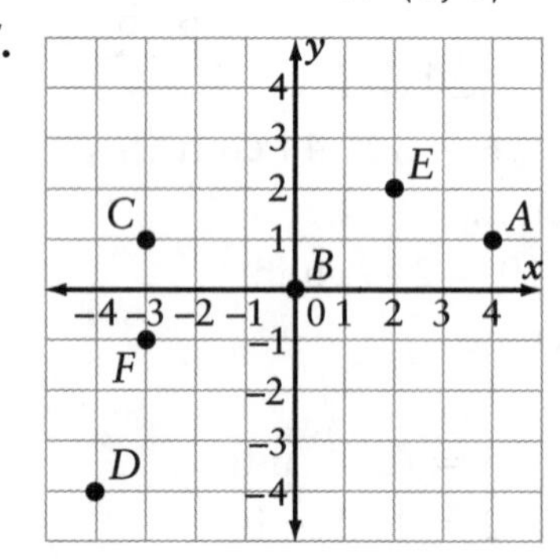

6.

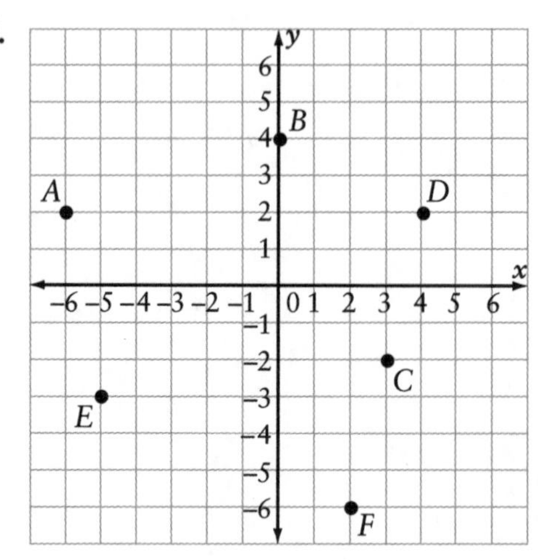

7.

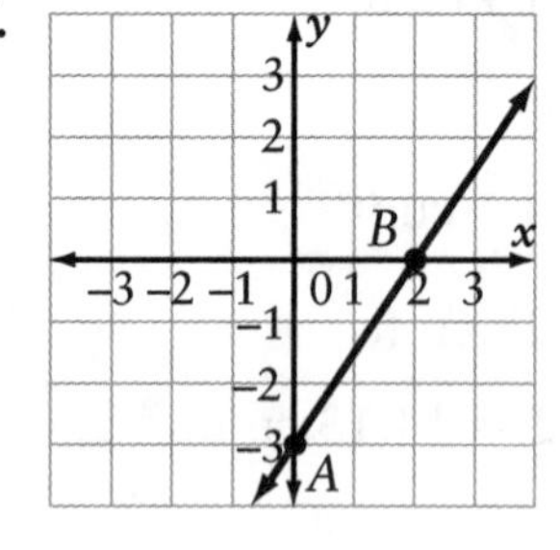

8.

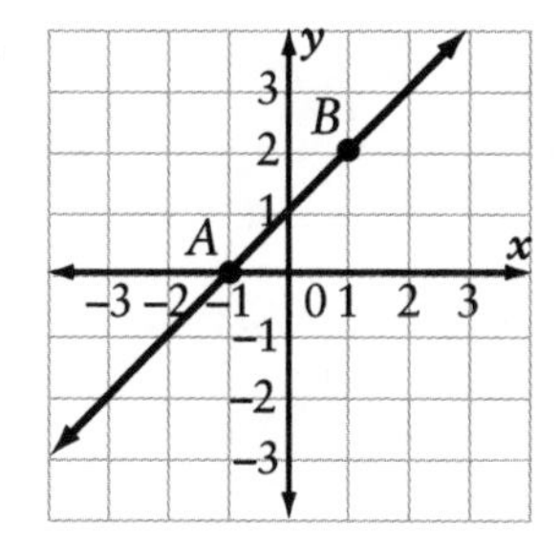

9.

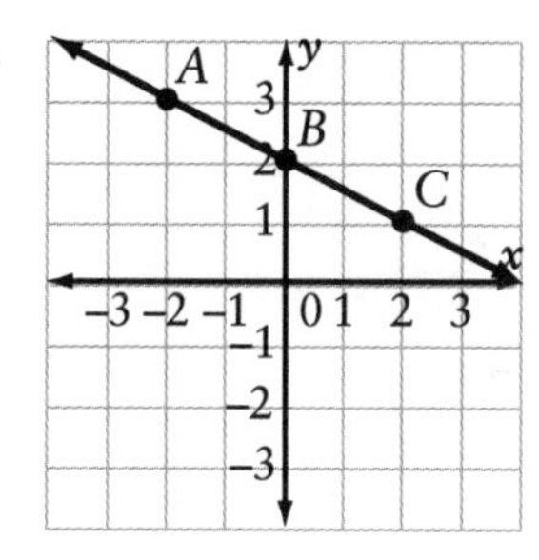

Coordinate Plane Review
page 102

1. origin

2. (0, 0)

3. (−3, −2)

4. (2, 0)

5. (−1, 3)

6. (3, −3)

Answers ***7–12*** on graph.

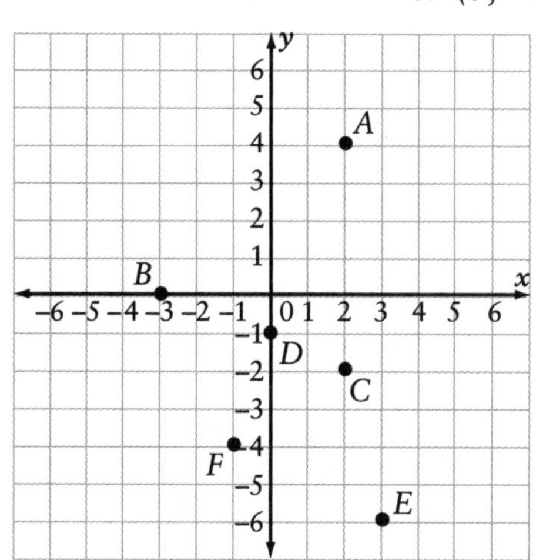

Final Review
pages 104–108

1. straight

2. obtuse

3. acute

4. reflex

5. right

6. $90 - 74 = 16°$

7. $180 - 105 = 75°$

8. transversal

9. $\angle 3$

10. parallel

11. $180° - 60° = 120°$

12. $\angle 2$

13. $180 - 112 = 68°$

14. equilateral

15. scalene

16. scalene

17. isosceles

18. $A = \frac{1}{2} \cdot 12 \cdot 8 = 48$ sq units

19. $P = 10 + 10 + 12 = 32$ units

20. 12

21. 6

22. rectangle

23. parallelogram

24. trapezoid

25. square

26. $A = 13 \times 7 = 91$ sq ft
$P = 2 \cdot 13 + 2 \cdot 7 = 40$ ft

27. $A = 15^2 = 225$ sq in
    $P = 4 \cdot 15 = 60$ in
28. $A = 31 \times 12 = 372$ sq m
    $P = 2 \cdot 31 + 2 \cdot 13 = 88$ m
29. $A = 3.14 \cdot 4^2 = 50.24$ sq in
    $C = 3.14 \cdot 2 \cdot 4 = 25.12$ in
30. rectangular solid
    $V = 10 \cdot 6 \cdot 8 = 480$ cu in
31. cylinder
    $V = 3.14 \times 5^2 \times 4 = 314$ cu ft
32. cube
    $V = 5^3 = 125$ cubic inches
33. $(-2, -4)$
34. $(-2, 1)$
35. $(1, -3)$
36. $(3, 2)$
37. $(0, 0)$
38. $P = 2 \cdot 60 + 2 \cdot 84 = 288'$
39. $60' = 20$ yd
    $84' = 28$ yd
    $A = 20 \cdot 28 = 560$ sq yd
    $560 \div 200 = 2+$
    Jeremy needs 4 bags.
40. $8'' = \frac{2}{3}'$
    $V = 10 \times 15 \times \frac{2}{3}$
    $V = 100$ cu ft